Village Homes' Solar House Designs

Village Homes' Solar House Designs

a collection of 43 energy-conscious house designs

by David Bainbridge
Judy Corbett
John Hofacre

with photographs by Carl Doney and Chad Ankele,
illustration assistance by Tonya Olsen, and
layout assistance by Carol Stickles

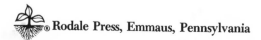 Rodale Press, Emmaus, Pennsylvania

Printed in the United States of America on recycled paper, containing a high percentage of de-inked fiber.

Library of Congress Cataloging in Publication Data

Bainbridge, David A
    Village Homes' solar house designs.

    Bibliography: p.
    1. Solar houses—Design and construction.
I. Corbett, Judy, joint author. II. Hofacre, John,
joint author. III. Title.
TH7414.B34   728.6'9   78–31780
ISBN 0-87857-261-9

  4 6 8 10 9 7 5 3

Dedication

This book is dedicated to Peter and Florence van Dresser in recognition of their support for this type of development over the last 30 years. In a small village in northern New Mexico, they have developed and refined the concept of regional self-sufficiency and have done it while living honestly with respect for the community and the environment. Peter's two books, *A Landscape for Humans* and *Homegrown Sundwellings* are both masterpieces and will hearten even the most discouraged cynic.

It is time we all followed their example and began thinking of our children's children's children rather than our own desires for today. Only by thinking about the future with a much broader perspective can we prevent increasing tragedies like Seveso, Italy; Minamata, Japan; the Manhattan tanker spill; the PPB contamination of food in Michigan; the radiation contamination of Bikini Atoll; etc.

Their approach is well illustrated by this quote from *Homegrown Sundwellings:*

There is still a widespread impression that solar heating is a complicated matter of large aluminum, copper, and glass structures, intricate piping and electrical controls which require an engineer to understand and operate and a financier to pay for. It is the main purpose of this book to correct this impression and show how solar energy can be harvested by simple means . . . well within the grasp of average homeowners and builders.

We concur.

Contents

Foreword

The design and successful development of Village Homes in Davis, California, represents far more than a significant milestone in the history of America's architecture and community design. What Michael and Judy Corbett have accomplished is a workable, no-nonsense model of a sensible, energy- and resource-conserving community of the future. Yet the future exists today, on the outskirts of Davis, in the Sacramento Valley.

I first became interested in the concept of integrated community design reading the works of Lewis Mumford, Paul Goodman and Murray Bookchin in the early 1960s. These writers have chronicled the waste of human as well as natural resources by poor urban design and the unwitting obeisance to many modern technologies—ranging from worship of the automobile and freeway to the gross inefficiency of the glass-walled skyscraper. I discovered the work of Sir Ebenezer Howard, who pro-posed the creation of "garden cities," in his prophetic volume, *Garden Cities of Tomorrow* (1902). Howard's proposed garden cities would incorporate fairly dense housing with wide agricultural belts, synthesizing the best advantages of the city as well as the countryside.

Howard's idea was to relieve urban congestion by dispersing the human population, and making the use of resources in small industries and farming more accessible. As Lewis Mumford points out, the garden city is *not* a suburb, "but the antithesis of a suburb: not a mere rural retreat, but a more integrated foundation for an effective urban life."

What Howard's pioneering work has done for the theory of realistic community design is surely matched by the Corbetts' concrete accomplishment at Davis. Village Homes is, in many respects, a modern model of a garden city, resting not on Howard's 6,000 acres, but on 70 acres. Central to the Village Homes' concept is the community's energy-conscious land use plan, which has resulted in the arrangement of houses in clusters with shared acreage controlled by their own homeowner's association. The overall design of this community is oriented towards people and their needs, not the automobile—which is the case in an average American suburb.

The streets are narrower than the norm, 20 to 25 ft. wide, rather than 32 ft., and considerable off-street parking is available for autos. Long cul-de-sacs as well as bikeways help to minimize the need for extensive automobile usage. Additionally, more than half the homes in the community utilize solar energy in various forms, ranging from striking, modernistic active systems to numerous passive designs, relying on climatically sensitive architecture.

The importance of Village Homes is underscored by the fact that this community is not a federally sponsored demonstration project of exotic technol-

9

ogy, nor is it a wild-eyed utopian dream. It is a realistic and commercially viable concept which has attracted many resident families by the logic and grace of its design.

An important lesson of Village Homes is that saving energy and resources need not involve sacrifice—it can be accomplished by careful design. Every home or apartment building, every office and skyscraper built in the United States has a predetermined energy appetite. According to a recent study of the American Institute of Architects, "a high-priority national program emphasizing energy-efficient buildings" would result in the savings of the equivalent of over 12 million barrels of petroleum per day—about two-thirds of what we now use daily in the United States. By reducing the energy appetite of new buildings, we can significantly reduce the need for new power plants and other energy facilities.

The addition of more planned communities of this sort will help to reduce energy and resources in a variety of ways, including the minimization of natural resource use for manufacturing and transportation. Less materials are needed for denser housing and narrower roads—resulting in substantial savings for energy-intensive materials such as concrete, steel and asphalt.

A persistent myth afoot in the country is that we can continue to grow indefinitely, and substitute the dwindling fossil fuels with other energy supplies. For some, the magic solution is nuclear; for others, it is solar power. But missing from such arguments is the necessity to deal with the realities of today's built environment. By reducing the extravagant use of materials and energy used for housing, transportation and manufacturing products, we can build an efficient and prosperous future through careful design. At the same time, capital needs for centralized energy facilities can be reduced, freeing funds for other elements of the economy.

Village Homes is a significant preview of what the future is all about. Today, some of America's great cities are crumbling into disrepair and bankruptcy, as the funds and materials are not available for their continuation. How ironic that in a suburb of Davis the answer to many of these woes is growing, not declining. Were Ebenezer Howard alive today, he would surely enjoy the success of the Village Homes experiment. As he wrote: "Town and country must be married, and out of this joyous union will spring a new hope, a new life, a new civilization."

Wilson Clark, energy and environmental adviser to California Governor Jerry Brown, and author of Energy for Survival, *and other books and articles on natural resources*

Acknowledgments

A book like this is made possible by the contributions of many people and we would like to give special thanks to those who were most instrumental in making this book a reality.

Mike Corbett, Village Homes designer, planner, and developer, for the energy, vision, and enthusiasm that brought Village Homes into existence.

The Village homeowners for their assistance and permission to describe their homes.

David Springer, for able assistance in writing the water heating and active space heating sections.

Lola Brocksen, for turning illegible scrawlings into beautifully typed copy.

Shelley Giacapuzzi, for help with the typing.

Bruce Melzer, for a very careful review and constructive comments.

Marshall Hunt, for the time to work on the book.

Robert Sommer, for sharing his knowledge, insight, and enthusiasm about our world.

Carol Hupping Stoner, for getting us started in earnest.

And to all our friends and family members who kept us sane throughout the process—many thanks!

Introduction

This book describes many of the solar homes in the Village Homes subdivision in Davis, California. It includes an introduction to the broader concepts of planning used in building Village Homes. A detailed description is given of the basic principles of building energy-efficient solar houses and the specific construction details for various components integral to these designs, including three types of solar water heaters. Eight solar house types are described and illustrated with floor plans, photos, and descriptions of 43 existing houses in Village Homes.

This book was written to fill what we three see as a gaping hole between public understanding of solar utilization and the actual state-of-the-art. Natural heating and cooling using solar energy and climate resources is not a complex technology—"good in the future, but not ready yet"—but a simple, easily applied, and well-tested approach to building that can improve the quality of all our lives and is ready now.

All three of us have been continually surprised by the amazement people show when they see Village Homes, and we decided that the many people who can't visit should also be able to understand what has been done here. This solar subdivision near Sacramento, California, currently has over 80 solar homes and will have over 200 when it is finished. It was developed with standard financing and has proven very successful as a commercial venture. The solar homes have been designed by a wide variety of people, predominately nonarchitects, and built by a number of local builders, both with and without previous solar experience.

From a modest beginning in 1972, when discussions and planning first began, to ground breaking in 1975, to the current level of development, it has been extremely interesting for us to be involved in the community and to watch it grow. The first houses were sold to young professionals who believed in both solar energy and saving money, but sales now are representative of the general market in Davis. Three years of experience in the development have proved that it works and is a good place to live. That is all anyone can ask.

The energy use of the average home in Village Homes is about half that of a comparable nonsolar home in the area.[1]* This remarkable saving can be attributed in large part to the careful use of energy-saving features on the naturally heated and cooled houses with solar hot water heaters. The rest of the savings result from minor changes in lifestyle that are an almost unconscious response to living with much lower energy requirements.

The savings are not at the cost of comfort or convenience either. In fact, many of the houses have proven much more comfortable than standard nonsolar houses. These improvements are not illusory; they can be explained by actual changes in the physical environment in these houses. This increase in comfort is perhaps the one thing that draws the most visitor comment, and it must be experienced to be believed.

We hope this book will encourage you to make changes in your present house or in plans for your house to be. To help you do so, we have provided a list of references and resources describing the various principles of micro-climate, human comfort, energy conservation, and solar energy that are involved in developing a solar development or solar home. Using solar energy is simple, economical, and can provide you with a more comfortable and secure home.

Best wishes on your solar future.

David Bainbridge
Judy Corbett
John Hofacre
  Village Homes
  Davis, California, 1979

*References are included at the end of the text.

13

Chapter One

Planning

The design of a subdivision or development is a tremendous responsibility. The choice of lot layout, street design, circulation patterns, and almost all other design considerations will affect people's satisfaction and sense of community, energy use, and resource use (water, minerals, etc.) for the next 50 years or more. This responsibility to the future has been ignored and abused in almost all subdivision design in favor of the "fast buck" and doing things the "way they are always done." Village Homes is a unique, courageous, and profitable exception.

This chapter describes the planning principles and practices used in this 70-acre development. It begins with a description of what is perhaps the most important aspect of the development: the methods used to help establish a sense of community. The next section explains the intensive land use in the development and the reasons for it.

Next is an examination of the energy conservation features incorporated in the planning of the development, features which have reduced energy use to only 57% of the energy consumed in a neighboring subdivision.[2] A look at special features used for resource conservation in the development follows. And the final section examines the success of the development to date and directions for the future, both at Village Homes and in the second generation of energy-efficient developments.

Community

A sense of community is an unknown in most of the recent suburbs and developments in America. Anonymous neighbors, fences, and commuter jobs encourage the development of alienation and anxiety that threatens to corrode our society and greatly diminishes the satisfaction in many people's lives. One of the dominant factors in the design of Village Homes was a realization of the importance of community, and many features were included to encourage the development of a strong sense of community.

To establish this sense of community, people must know their neighbors, and they will get to know them only if they have reasons to get together. Village Homes has made getting together easy and essential by setting up common areas of greenbelts, which are controlled by eight families, who were in most cases involved from design through to construction. After completion most of the maintenance is also done by the cluster members. This has not always been easy for those of us unused to sharing responsibility, but it has been very effective in establishing community.

Working together on community projects has also been encouraged, both to reduce costs and let people get to

know each other. Work parties have been held to build retaining walls, bridges, play areas, the pool complex, and community center. This has the added benefit of giving people the pride of ownership and has resulted in much better care and protection of community projects.

Community is also encouraged by giving people reasons to be outside so they can interact. The use of small private yards on the street side with large open yards in the back has made gardening very popular. The lack of fences and the circulation of pedestrians and bicyclists on the greenbelt path allows for much needed interaction and the never-ending discussions, popular with gardeners everywhere, about the weather, crops, bugs, etc.

Too much through traffic can disrupt and eventually destroy a community. In Village Homes, long cul-de-sacs were used to eliminate through traffic. As the residents of each cul-de-sac get to know each other they can maintain careful watch on the street. This "defensible space," as Oscar Newman[3] calls it, is very important for all of us.

Finally, a community is weakened

Neighbors join in to build community facilities at a work party.

if most or all of the employment opportunities are outside the neighborhood.
Village Homes includes several features
to provide as many jobs in the community as possible. These include setting
aside commercial space for community
job development, agricultural areas for
community farmers, and the greenbelts
and community facilities that are all
maintained by members of the community and provide considerable income.
In addition, some of the residents work
at home. Much more should be done to
encourage professional workers to work
at home most of the week.

We have found it very satisfying to
live in a community where on an evening walk you can see and talk to your
neighbors as they putter in their garden, or walk and visit friends who are
close and accessible, in large part because of the careful foresight in design
and intensive land use in Village Homes
described in detail in the next section.

Intensive Land Use

Land has typically been squan-
dered as if it had no value, except what
it would bring on the market on the day
of sale. The design of Village Homes
was based on a clear understanding of
the value of land as a resource, particularly the rich agricultural land in Davis.

An examination of land use in a
standard development will show why
major changes in land use were made in
Village Homes. The land use allocation

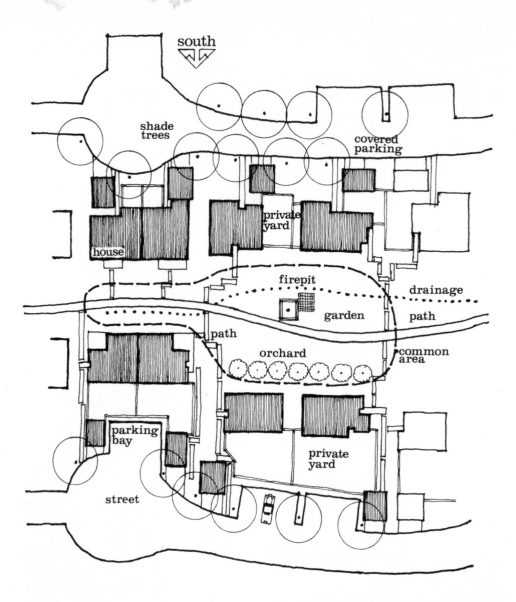

16

in a standard development in Davis is shown in the following chart.

LAND USE ALLOCATION IN A TYPICAL DEVELOPMENT[4]

Land Use Type	Percent Total
Backyard	22
Front yard	19
Street (40 ft.)	18
Houses	15
Side yards	11
Garages	5
Sidewalks	5
Driveways	5

This misuse of land is no longer viable. Almost 30% of the development is under pavement, only 20% is under structures, and only 20% (the backyard) is really used much, and typically less than 5% is used productively in a small backyard garden. Every one of us will pay the price for this misuse of resources in the future as we try to improve efficiency of our developments to survive in a resource-scarce world.

Village Homes was designed for intensive and efficient land use. Lot sizes were reduced by making front yards much smaller and turning them into private space which is used rather than maintained at considerable cost for "show." Side yards were also reduced by building some commonwall houses and reducing side yard setbacks. And rear yard size was reduced by opening all the backyards as a common greenbelt to

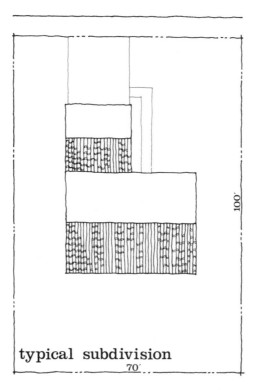

street 32'

typical subdivision

give the desired feeling of space. This openness also makes bigger gardens more feasible and more common. The smaller lots have also encouraged people to build more compact houses and two-story houses.

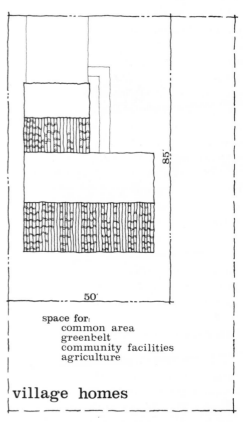

street 25'

space for:
common area
greenbelt
community facilities
agriculture

village homes

Land use was also improved by reducing street widths from 32 ft. to 20-25 ft., by using parking bays for parking, and also by having a sidewalk that serves as a bikeway running down one centrally located greenbelt rather than

17

LAND USE ALLOCATION IN VILLAGE HOMES
vs. STANDARD DEVELOPMENT

Land Use Type	Village Homes Percent Total	Standard Development Percent Total	Percent Change
Backyard	12	22	− 10
Front yard	10	19	− 9
Street (including parking bays)	14	18	− 3
Houses	12	15	− 4
Side yards	4	11	− 7
Garages and carports	5	5	0
Sidewalks and bikeways	8	5	+ 3
Driveways	3	5	− 2
Agricultural area	17	0	+17
Common area	15	0	+15

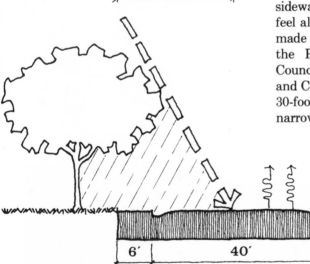

26′

6′   40′   6′

street shading

sidewalks on both sides of the street. We feel all streets could have actually been made narrower with equal success, but the Planning Commission and City Council, based on the Planning Director and City Engineer's recommendation of 30-foot streets, were not willing to go narrower than 20-25 ft. The main con-

cern of the city officials was fire access, but a recent study suggests this is not a valid concern.[5] Streets were laid out as cul-de-sacs without connecting them on a grid, saving even more pavement. Many two-story homes were used, also saving land.

Adding all these factors together frees land for other uses. Most of this liberated land in Village Homes was used to create a community farm of 15 acres which includes orchards, vineyards, and vegetable fields. This saves a valuable resource from destruction and also provides many benefits to the community. Taken together, these changes make a significant impact on land use as shown in the chart comparing land use allocation in Village Homes and a standard development.

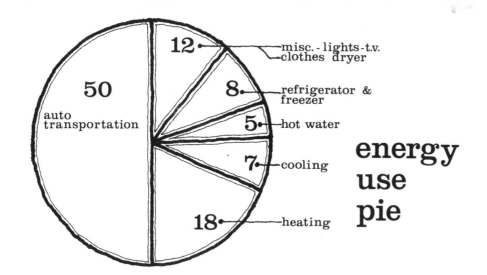

energy use pie

- 50 — auto transportation
- 12 — misc.-lights-t.v. / clothes dryer
- 8 — refrigerator & freezer
- 5 — hot water
- 7 — cooling
- 18 — heating

To discuss the importance of energy conservation features we need to know how energy is used in Davis by a typical family. The energy pie included here shows how energy was used in Davis when the most recent study was done.[8] The major use of energy is for transportation; this represents over one-half of the total. Village Homes and the city of Davis have both taken the first steps to reduce energy used for transportation, but unfortunately local impact is limited at best and only modest savings can be realized. A look at the chart of Net Energy Use for Transpor-

These changes in land use not only make the development more efficient, as we shall see in the next section on energy conservation, but also make it a much more pleasant place to live and reduce the cost of the development.

Energy Conservation and the Use of Natural Energy Sources

Village Homes and the city of Davis are perhaps best known for their energy conservation efforts, although as we have tried to show, energy conservation is only one aspect of the development.[6,7] Although the energy conservation effort was one of the easiest and simplest parts of the design, it has had a major impact on energy use and holds an important lesson for other areas.

Bikeways take the place of roads in parts of Village Homes.

tation shows very clearly which transportation modes we should be encouraging:

NET ENERGY USE
FOR TRANSPORTATION
PER PASSENGER MILE[9]

	Btu
Walking	800
Bicycle	1,400
Regular bus	5,000
Automobile	10,000

Both Village Homes and, to a lesser extent, the city of Davis have made a concerted effort to promote walking and bicycling. Bicycles have received the most attention with encouraging progress. In 1976 there were 33,000 people and 25,000 bicycles which were ridden over a million miles in the city of Davis.

Bicyclists and walkers are encouraged in Village Homes by the separate bike and pedestrian paths (not on roads), bike parking areas, and the circulation pattern. It was deliberately made easier and faster to walk from one area to another in the development than to drive there. And the whole network is tied into the city bikeway network. Numerous recreational facilities and the provision of jobs for members of the community in a commercial center and in agricultural projects will reduce the dependence of residents on the automobile.

The second largest use of energy is for space heating and cooling, and this is the area where Village Homes has been most successful in achieving savings. The following chapters describe the change in building design used to achieve these savings. The planning of the neighborhood was designed to facilitate solar heating and natural cooling, and the very simple steps necessary for solar utilization were among the first made and most vigorously pursued. These include: street orientation, lot orientation, setbacks, solar access, and landscaping.

Street orientation is probably the most important step. Streets run predominately east-west to ensure that houses have their major walls on the south and minimal exposure to the east and west. By itself this "sun tempering" can reduce energy use for heating and cooling 20-50%. This is mandated now by the Davis Energy Conservation Building Code,[10] but it was undertaken in Village Homes before the Code was implemented.

The second modification used to reduce energy use for heating and cooling was the modification of lot shape to make lot sidelines run north-south even on the gently curving streets. This minor and valuable change was unacceptable to FHA officials who refused to loan for the houses unless lots were shaped more traditionally, which means running perpendicular to the streets.

The third technique used to minimize energy use for heating and cooling involved changes in street-side setback for front, back, and side yards. Houses are set comparatively further back from the street when a courtyard is needed to enclose extensive south glass and give privacy. A fence may be as close to the street as 10 ft. or as close to a parking bay as 2 ft. The house may be placed within 5 ft. of the back property line. Commonwall construction has been used occasionally and has been successful in saving land, thereby reducing the cost of development. It also reduces unwanted summer heat gain on the east and west by eliminating east and west walls; it reduces winter heat loss as well.

The fourth step involved protection of solar access and essentially established solar rights for residents of the development. Through the design review process and the covenants of the development, exposure to the sun is carefully protected. Before construction begins, a model of each house is made and placed on a large plan of the development. An architectural review board makes certain that buildings do not shade one another. Shading due to vegetation is prevented by a special amendment to the Homeowner's Covenants Codes and Restrictions (Appendix A).

The next change involves better use of landscaping material. Trees, for example, are carefully chosen for maximum shade in the summer with fast leaf drop in the fall and minimum branching to reduce shading in the winter. Winter exposure is also protected by placing trees predominately on the east and west of houses where their shade is most helpful for cooling.

The openness of the backyards in the development also aids in cooling by permitting good cross-ventilation from the cool evening sea breezes. And finally, the narrow streets and parking bays can be more thoroughly shaded than a wide street, and summer heating can be greatly reduced. When trees are more fully grown, a 10°F reduction in air temperature is expected on hot summer days.[11]

Water heating is the next most important energy use in Davis, and the protection of solar exposure for space heating also protects solar water heating from shading. Most of the houses are on full solar water heating for seven months of the year and 60–80% solar water heating during the rest of the year.

And finally, although it doesn't appear on the energy use pie, the use of energy for food production is reduced by the emphasis on local gardens. Food products channeled through the traditional market cost about eight calories of energy for each calorie of food while those grown in home gardens may cost only 0.5 calories per calorie of food value.[12] The residents also get the benefit of tastier and more nutritious food and minimize the use of the many resources needed to keep the refrigerator trucks, trains, and tractors rolling.

Resource Use

Village Homes has also taken many other steps to reduce the use of other resources for living. These include water conservation measures, use of greywater (although this has created conflict with the County Health Department officials), composting, recycling, and use of durable materials for construction. Sadly, one of the few failures in the development of Village Homes was the inability of the developer to get permission for innovative sewage treatment using composting toilets and cluster septic tanks. Although the ideas were sound and the concepts are field-tested in the United States and widely used in Europe, it was simply one change too many to carry and was ultimately dropped.

Compost was originally collected free of charge by Village gardeners, composted in centrally located drum composters and used on community lands. It soon became apparent that those who had vegetable gardens wanted to keep their compost for their own use. In addition, the drums proved less efficient than simple compost heaps. Composting is now done individually,

Architecture Review Board members study solar rights with models.

except for a community area where non-vegetable gardeners dump ornamental garden trimmings for composting.

Efforts to reduce water consumption include: aboveground drainage to save rainfall, use of drought-tolerant landscaping, and installation of water conservation devices in the houses. All of these, except aboveground drainage, are now standard practice throughout Davis. The drainage runs through the greenbelts rather than storm sewers, and these greenbelts handled even 150% of normal rainfall without difficulty with the release of only 20% of the rainfall to the city flood-control system. This water is stored in the shallow aquifer for use by the trees and to a lesser extent by the shrubs.

And finally, materials for houses at Village Homes were chosen for durability. A couple of examples will suffice. The roof tile used on most houses is a concrete tile with a lifetime well over 100 years so that the energy cost is very low over the life of the development. The use of concrete for the parking bays and bicycle paths is more expensive, yet will have a lifetime much longer than asphalt.

In Summation

Taken together, these many changes provide substantial improvements in the development of the earth's limited resources of energy and materials. The cost of the development has been comparable to a standard development, with some changes costing more and some considerably less than standard practice. As they say, the "proof is in the pudding," and the satisfaction of the members of the community is the real proof of the validity of the concept and the success of the development.

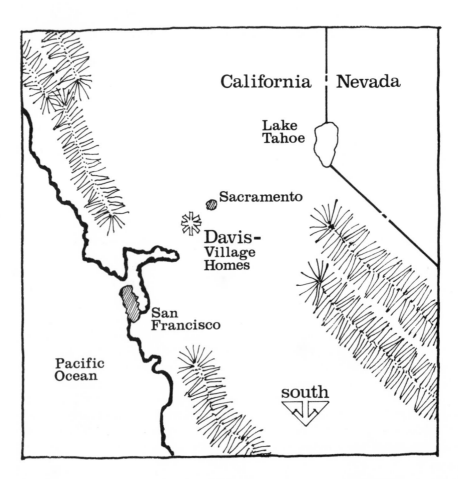

Chapter Two

Concepts and Details

Although the homes in Village Homes use very different types of solar systems, they all share many concepts and details of design and construction. These include energy conservation features such as insulation and weather stripping, solar orientation, window placement for cross-ventilation, extra mass, durable materials, and energy-efficient appliances. Some of these are specifically designed to fit the local climate but most of them are applicable almost everywhere.

The climate in Davis is known as a Mediterranean climate. This describes a fairly temperate climate with strong marine influence. The summer days are hot (reaching 100°F fairly regularly), but as the hot air rises over the Sacramento Valley it draws in the cool air from the Pacific Ocean and nighttime lows are comfortably cool (averaging about 56°F over the summer). The Pacific also moderates the winter ex-

tremes, and freezing temperatures are a fairly rare occurrence.

Heating and cooling degree-days have traditionally been used to describe the need for heating and cooling. Although the degree-day is not ideal, particularly for solar design, it is widely used, and degree-day data has been compiled throughout the United States. A heating degree-day is one degree below the desired temperature, 65°F, and a cooling degree-day is the opposite, or one day in which the temperature is one degree over the desired temperature, usually 65°F. The choice of 65°F for both is unfortunate, but it is commonly used in heating and ventilation design. If 65°F is used as the base for both heating and cooling degree-days, Davis has 2,800 heating degree-days and 1,100 cooling degree-days. This suggests that house design should pay almost equal attention to both solar heating and natural cooling. However, the first and

most economical step in any house design is careful consideration of the opportunities for energy conservation, including insulation, weatherproofing, and careful choice of materials.

Energy Conservation

Careful insulation is one of the first steps in reducing energy demand in a house. Walls and roofs are almost always insulated now, but thicker insulation is often desirable. Many of the houses in Village Homes have used 2×6 stud walls instead of 2×4 walls so that more insulation can be put between inner and outer walls to increase their houses' "R" value, or resistance to heat flow, from R-11 to R-19. Roof insulation is typically R-30, compared to R-19 in standard houses nearby.

Insulating the other exposed areas of a house is also important and often not done. For example, the edge of the floor slab is often not insulated and as a

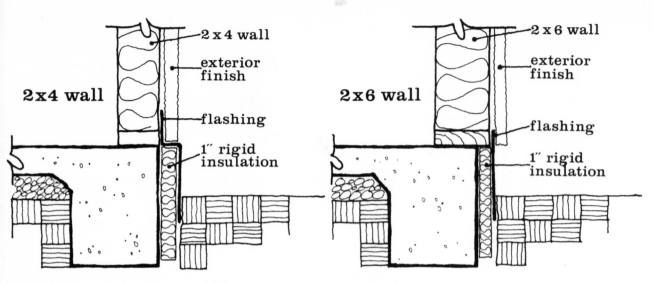

2x4 wall

- 2 x 4 wall
- exterior finish
- flashing
- 1″ rigid insulation

2x6 wall

- 2 x 6 wall
- exterior finish
- flashing
- 1″ rigid insulation

slab insulation

result the slab provides a path for unwanted heat loss in the winter and heat gain in the summer. It is inexpensive and easy to add insulation to the edge of the slab during construction, and almost all of the builders in Village Homes do so as a standard practice.

Windows are another area where energy conservation must be considered. Although the primary concern is usually radiation gain from sunshine (as discussed later in this chapter), windows are also an important factor in conductive heat loss in the winter and heat gain in the summer. A single pane window, for example, has an R-value of only 0.9 or only 1/20 that of the typical wall used in Village Homes. The use of

double or thermal pane windows increases the R-value to 1.8 which is better, although still not very good.

As a result, many of the houses in Village Homes further improve their windows by providing thermal shutters or drapes. These come in many types with widely different costs and vary from R-values of only 2–3 up to 19. A good thermal drape or shutter should be a good insulator, able to seal tightly at the edges, bottom, and top; durable; easy to operate; and fire-retardant.

Three examples are shown here. The first example is an insulated shutter built using mahogany door skin, fiberglass insulation, and a wood frame. For smaller windows or even large win-

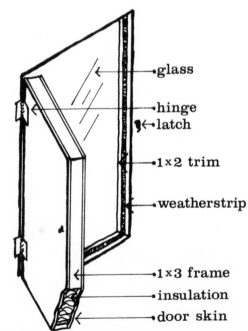

- glass
- hinge
- latch
- 1 x 2 trim
- weatherstrip
- 1 x 3 frame
- insulation
- door skin

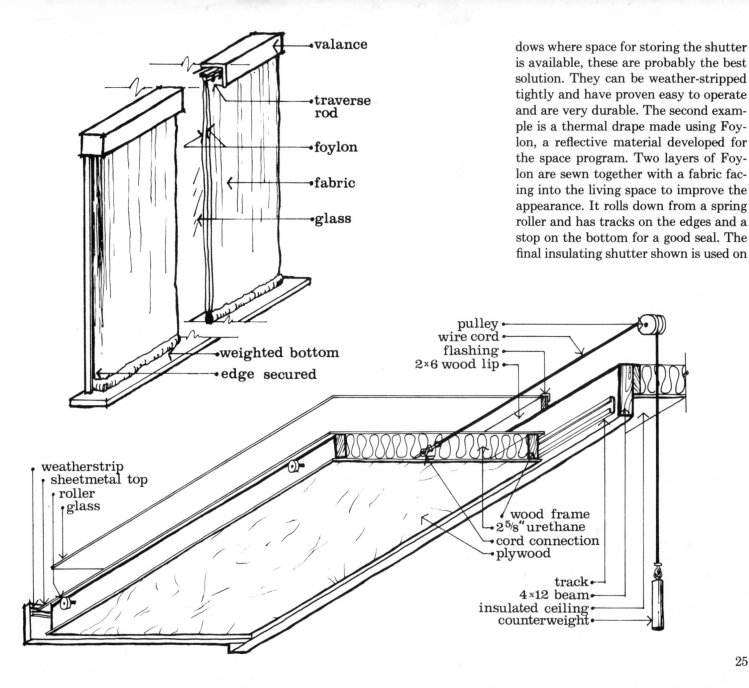

valance

traverse rod

foylon

fabric

glass

weighted bottom

edge secured

dows where space for storing the shutter is available, these are probably the best solution. They can be weather-stripped tightly and have proven easy to operate and are very durable. The second example is a thermal drape made using Foylon, a reflective material developed for the space program. Two layers of Foylon are sewn together with a fabric facing into the living space to improve the appearance. It rolls down from a spring roller and has tracks on the edges and a stop on the bottom for a good seal. The final insulating shutter shown is used on

pulley

wire cord

flashing

2×6 wood lip

weatherstrip

sheetmetal top

roller

glass

wood frame

2⅝" urethane

cord connection

plywood

track

4×12 beam

insulated ceiling

counterweight

the roof skylights used in Village Homes. It is the most expensive and complicated shutter used in the Village, but it has to be to provide the thermal control needed.

Another facet of energy conserving design is the choice of exterior color for walls and roof, and the choice of roofing and siding material. Unwanted heat gain can be greatly reduced by using light-colored walls, and most houses in Village Homes use white or off-white stucco for this reason.

The choice of roof color and materials also affects heat gain. Village Homes allows only tile and shakes because they proved to have the best thermal performance in tests carried out near Davis. Shakes and concrete tile remained the coolest of the six roofing materials tested.[13] Light-colored tile and shakes remained 40°F below light-colored metal roofing or composition roofing on a clear sunny summer day. A wide variety of concrete tile colors has been used in the development. Even the darker tile performs better than most traditional roofing materials because it is well ventilated.

Weatherproofing

The final energy conservation measure involves tightening up the house, or weatherproofing it. This involves carefully sealing the many areas where unwanted cold (or hot) air can enter the house. This is important because in a well-insulated house infiltration losses may almost equal conductive losses.[14]

The construction of the house must be done with an eye to reducing infiltration. The sills should be set in mastic to reduce leakage under the walls. This is done with almost all the houses built in Village Homes and it makes a big difference. Caulking should also be done around doors and windows where cracks are left between the rough frame and the finished window or door frame.

The choice of windows and doors is equally important. The windows used in Village Homes are expensive, but are very well built and seal tightly. Doors can be weather-stripped many ways but the most durable and effective method uses interlocking metal pieces set in the door and door frame. This is shown in the accompanying detail.

Appliances

And finally, the choice of appliances will affect the performance of the house. A poor refrigerator may use 200 kwh per month, and this can add 600,000 unwanted Btu's of heat to the house in the summer. Although it will help heat the house in the winter, using the refrigerator to heat the house makes about as much sense as burning down the house to stay warmer. An efficient refrigerator may use only one-fourth as much and will save twice as much, both by reducing energy use and by reducing unwanted heat gain. It also helps to do without appliances that are not necessary, and many of the residents in Vil-

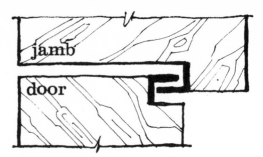

interlock weatherstrip

lage Homes have chosen to live without dishwashers, clothes dryers, garbage compactors, and other appliances which often provide very limited benefits at very high cost both in money and energy to purchase and operate. Most of the residents in Village Homes have also chosen solar water heaters.

Water Heating

Solar water heaters are reliable, practical, and economical. They have been used in many areas of the world for over 80 years. In California and Florida, a solar water heating industry developed in the early 1900s and by 1930 tens of thousands of units were in use.[15] These were gradually replaced by electric and gas heaters as energy subsidies by the government were increased.

Solar water heating makes sense. It can provide lower cost hot water and increase one's feeling of self-sufficiency. In addition, it helps conserve needed high-grade energy for more important uses. As Amory Lovins has said, "Heating water with electricity is like cutting butter with a chainsaw."

Of course, the solar system should not be installed or designed until all reasonable water conservation measures have been applied and the backup water heater tuned up. These steps can save a great deal of money even for people who never get the solar system of their dreams built.

Water conservation measures can usually reduce hot water demand by half or more.[16] These include such inexpensive and readily available equipment like constrictors on the shower and water faucet, pressure reducers, aerators, better pipe insulation, and more expensive ones such as higher-efficiency appliances, air injection showers, and heat exchangers. Most of these look extremely good when life cycle costs are considered. Strict application of all possible conservation measures can reduce hot water consumption to 10% of existing use.

Tune-up of the backup water heater includes retrofitting increased insulation, turning the thermostat down (140°F or less, not 160°F), cleaning the tank regularly, and installing a retrofit electronic ignition package if you can find one for your model.

There are three major types of solar water heaters used in Village Homes. In order of increasing complexity and cost they are the breadbox, thermosiphon, and pumped solar water heaters.

Breadbox

The breadbox hot water heater was the earliest solar hot water heater. The first major production followed the patent of the Climax water heater in 1891 and the "Walker" model in 1898. A detailed evaluation was done in 1936 by F. A. Brooks at the University of California[17] and then the breadbox was virtually forgotten until renewed activity began in 1974 both in Davis and elsewhere.[18] A breadbox water heater is simplicity itself — a black container of water set in an insulated box facing south. It requires no pumps or controls because it is plumbed in between the normal cold water intake and the hot water tank. As hot water is used, cold water is drawn in the intake side of the breadbox

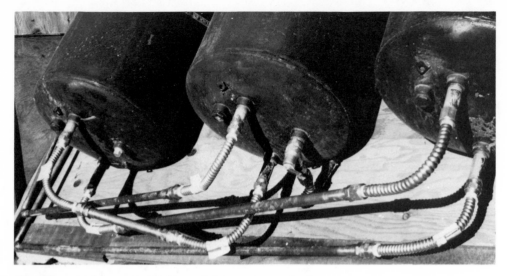

Plumbing provides water exchange between storage tanks.

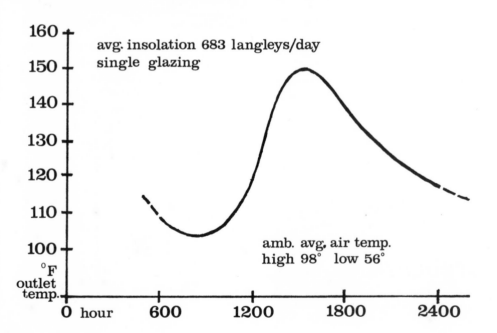

avg. insolation 683 langleys/day
single glazing

amb. avg. air temp.
high 98° low 56°

and starts warming up.

Temperature of delivered water is comparable to that provided by flat plate collectors. As the graph here shows, temperatures reach 150°F in the afternoon and drop to about 100°F in the early morning with a single glazed breadbox in Davis in the summer. Using double glazing, insulated pipe runs, and movable insulation at night, similar performance could probably be realized in most of the United States for a minimum of six months of the year. In mild areas like Davis, hot water demand has been fully provided for nine months with this type breadbox. And in warmer areas, like Southern California, Arizona, and the Gulf states, year-round needs may be met with these simple solar heaters.

The simplicity is an attractive feature because it makes the breadbox a solar heater almost anyone can build. The materials required are readily available, low in cost, and easy to put together with only limited knowledge of carpentry and plumbing.

The cost of materials for the three tank system shown here was around $400 in 1978, using new materials. Most parts can usually be scrounged for free or very low cost, and breadboxes have been built for as little as $30. Neither amount includes labor, and if one had to pay for labor at a cost of $10 per hour, a breadbox would cost about $400 more since it takes approximately 40 hours to assemble.

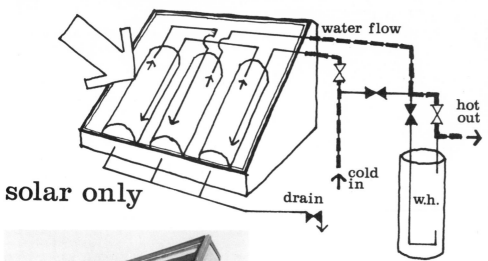

solar only

water flow

hot out

cold in

drain

w.h.

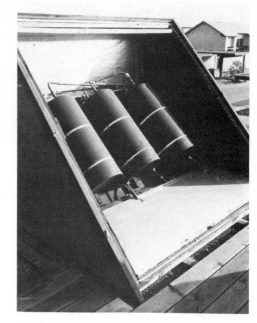

Tanks are secured in a "breadbox" on roof.

Pipes feed cold water to tanks and return hot water to house.

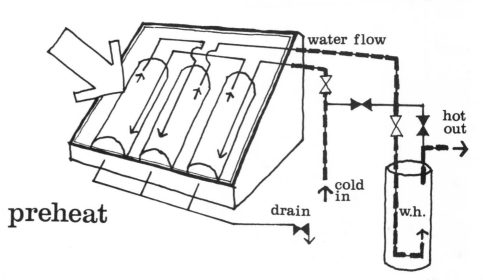

preheat

water flow

hot out

cold in

drain

w.h.

Pipes as well as hot water storage tanks are well insulated.

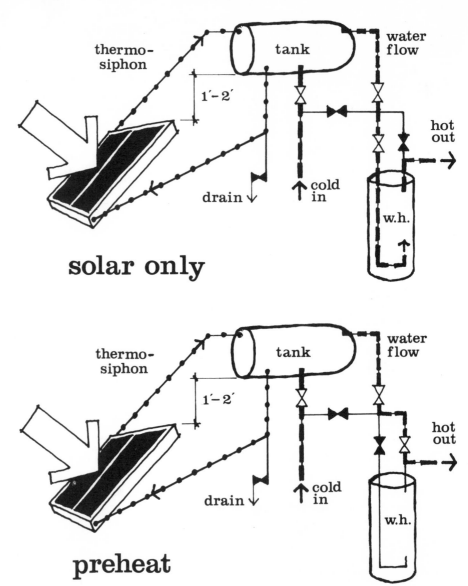

thermo-siphon

tank

water flow

1′–2′

hot out

drain

cold in

w.h.

solar only

thermo-siphon

tank

water flow

1′–2′

hot out

drain

cold in

w.h.

preheat

Close-up of roof-top solar panels.

Thermosiphon

In some cases, the flat plate collector may be easier to retrofit or install than the breadbox and can provide better performance in colder periods without the user having to move lids and shutters.

Water is warmed in the flat plate collector and rises (warm water is lighter than cold water) to a storage tank mounted above the collector. Because this water is stored in an insulated tank, it will stay warmer all night than an unshuttered breadbox solar collector. The use of the natural thermosiphon for circulation makes the use of pumps or controls unnecessary.

The thermosiphon system provides automatic freeze protection. If the storage tank is kept within one foot above the collector, then a reverse thermosiphon will occur, warming the collector.

The system design and orientation will determine how much water the thermosiphon system can provide. The basic size of collector and storage for homes in Village Homes was determined by assuming water use of 20-30 gallons per person. A collector will produce about one gallon per day per 1/2 to 3/4 square foot of collector.[19] This factor multiplied by the demand is used to give a fairly good estimate of required collector area. And storage of 1 1/2-2 gallons per square foot of collector is usually enough. A family of four using 80 gallons of hot water per day would typically require 40-60 sq. ft. of collector and an 80-120 gallon storage tank. Most of the systems in Village Homes fall in this range.

The systems used in Village Homes have been developed over four years of operation by Dave Springer and the Natural Heating System crew. With over 40 thermosiphon systems operating in Village Homes they have some 60 system years of operating experience, and it has enabled them to design, build, and install very reliable and efficient systems at a reasonable price.

The cost of materials for a typical thermosiphon system is around $800. Installed cost in Village Homes has been typically $1,200-1,800, although it is now up to $2,100. While not as simple to build as a breadbox water heater, the thermosiphon system can be built and installed by a self-educated homeowner. Attending a solar water heating workshop will often prove very useful and save a great deal of frustration and loss of time and money.

Pumped System

Space limitations may prevent the installation of either a breadbox or thermosiphon system, and then a pumped system must be used. The pumped system has also been in use for many years around the world. It is more complex in design and function and includes a flat plate collector, storage tank, pumps, valves, and differential thermostats which turn the pump on when the collector is hotter than the storage tank.

The pumped system is used when site conditions or the roof design do not allow for a thermosiphon tank above the collectors. The storage tank may be located anywhere but the closer to the collectors the better, reducing heat loss and the expense of long pipe runs. A pump size of 1/8 hp usually provides enough flow. Single pane glass has been used on the collectors. Double pane glass would reduce the heat loss, thus increasing efficiency; however, the

solar only

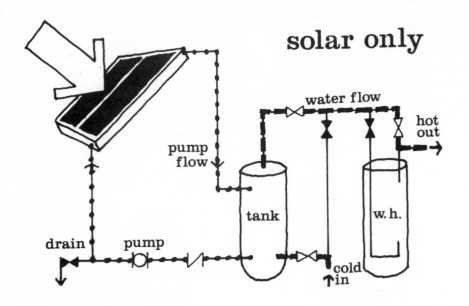

water flow

pump
flow

hot
out

tank

w. h.

drain

pump

cold
in

preheat

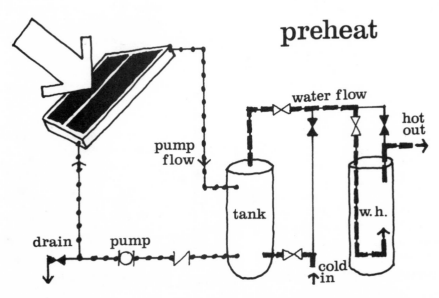

water flow

pump
flow

hot
out

tank

w. h.

drain

pump

cold
in

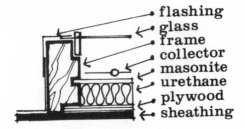

flashing
glass
frame
collector
masonite
urethane
plywood
sheathing

collector

added glass increases the initial reflection of light which in turn reduces the efficiency.

The amount of water that can be heated and stored again depends on many aspects of the system design, but the basic sizing rule given for the thermosiphon will generally prove adequate for the pumped system as well.

The performance of a pumped system will be similar to that for the thermosiphon system. Such systems have not been used often in Village Homes because of their very high price ($2,500 for a typical installation) and increased complexity. To eliminate problems with freezing, a two-tank system with antifreeze solution and double wall heat exchanger are used at an added cost.

Installation and design are more difficult and they are perhaps best left to a professional or very skilled handyperson. If one is built by a homeowner, we recommend a design check by an experienced installer/designer.

Components of pumped, active solar hot water system.

movement and humidity. Passive houses are more comfortable than standard houses over a broader range of air temperature because they provide very uniform radiant temperatures which are above air temperature in the winter and below air temperature in the summer. The increased comfort in passive houses is also in part due to better humidity balance, and the lack of low level irritation from noise, vibration, and blowing air.

One of the best things about passive solar systems is their simplicity. Anyone can design a workable passive solar system once they understand the basic principles involved. More sophisticated techniques, such as computer simulations, are primarily needed to figure out more precisely how the system

The Basics of Passive Solar Design

After all of these energy conservation features and considerations have been included in design, it is time to turn to another design consideration, passive solar. The first step is to explore the basics of human comfort, which is after all what we are about.

Comfort at normal room temperatures is about equally based on the air temperature, the surface temperatures of walls, floor and ceilings, and air

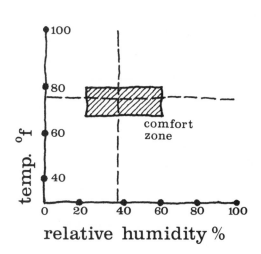

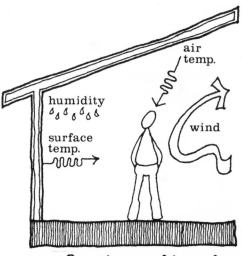

comfort criteria

will work, and such things as how much mass is optimal. But the relatively simple techniques and rules described here will work well. In fact, most of the homes built in Village Homes were built without extensive studies or detailed calculations.

The basics of passive solar heating and cooling include:

the sun and proper orientation
heat transfer
heat storage
ventilation and convective
    cooling.

The Sun

To use solar energy effectively, we must understand the sun's path and the nature of solar radiation. The sun's position is described by its height above the horizon, altitude, and its bearing from true north, azimuth. The change in elevation and azimuth over the season is a critical aspect of passive design. In Davis at about 38.5° north latitude, the sun is at an altitude of almost 30° at noon on December 21 and reaches an altitude of almost 70° during the summer. In summer it traverses an arc of 240° from east to west, while in winter it only covers 120°.

Solar energy is available almost every day of the year. It reaches the earth as either direct sun (predominant), diffuse sun (reflected by clouds and dust), or reflected sun (bounced off bright surfaces such as concrete, white roofing, or other bright surfaces). Passive systems

can utilize all three of these while many active systems work only with direct sun. The solar energy striking a typical house's south wall on a winter day contains as much energy as 11 gallons of gasoline. Needless to say, we only need to capture a portion of this energy to keep warm.

Proper orientation will maximize heat gain in the winter while minimizing unwanted heat gain in the summer. It does so by utilizing the difference in the sun's path over the year. A house that is longer east-west than north-south with most windows on the south and a modest overhang on the south will receive full sun on December 21, the

winter solstice, yet be fully shaded on June 21, the summer solstice.

The homes in Village Homes make full use of this principle. Almost all of them have very few windows on the east and west sides where summer heat gain is worst. The south side, in contrast, has many windows with overhangs or arbors to shade them in the summer. Arbors covered with deciduous vines are attractive because they are able to handle the swing seasons of spring and fall better than those without such vines. Cool springs delay leaves coming out and allow almost full sun, while a warm fall will keep leaves on the vines, providing needed shade.

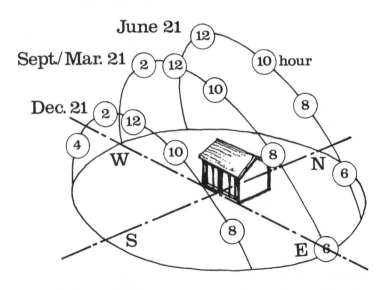

sun path

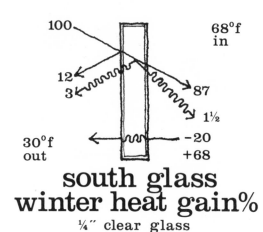

south glass
winter heat gain%
¼″ clear glass

Natural Cooling

Proper orientation also is desirable for natural cooling. The cool marine air comes through the Carquinez Straits and curls north to Davis. This south wind can provide very quick cooling in the evening if operable windows are included on the north and south sides of the house.

This natural ventilation can be complemented by designs that induce ventilation by using high and low vents. A high vent near the peak with an insulated and tightly sealed cover combined with low vents on the north side is particularly effective. Turbo vents can also be used.

A mechanical fan to the attic and a security vent can provide needed ventilation while the owner is away. The fan can be fairly small, around 1/10 hp and 500 cfm, and should be set on a timer circuit. A local manufacturer is marketing sliding glass doors which can be secured while still slightly open. Second-story windows are an additional type of security ventilation used in Village Homes. Other options are possible.

A "casablanca" type ceiling fan in the house can provide moving air for comfort on very hot days. Even a very light breeze will provide comfort if the air temperature is only slightly above the comfort zone.

Thermal Storage

The thermal storage possible in a house is also important, and the ability to store heat is determined by the thermal mass of a building. The massive adobe walls of the old missions are a classic example of thermal mass. Passive homes in Village Homes store energy using barrel walls, water tanks, ceramic tile, masonry, concrete, adobe or earth berms to store "heat" and "coolth" to bring the temperature extremes within the comfort zone.

It is worth a closer look at the properties of materials that have been used for thermal storage in Village Homes.

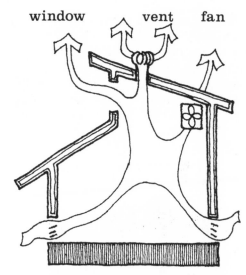

ventilation

Water storage costs about one-fourth as much as concrete for a given amount of thermal storage. Water is also more effective than concrete in soaking up energy because it circulates. This is particularly important for natural cooling. The chart of water storage containers includes relevant informa-

PROPERTIES OF THERMAL STORAGE MATERIAL

Material	Specific Heat (Btu/lb/°F)	Density (lb/ft³)	Heat Capacity (Btu/ft³/°F)
Water	1.00	62	62
Concrete	0.23	140	32
Stone	0.21	170	36
Brick, tile	0.21	140	28

Empty culverts waiting to be moved to houses where they will be filled with water to provide thermal mass.

Water-filled barrels add to the heat-storage capacity of the house.

tion on the different properties of the containers.

Concrete is expensive if used only for storage, but if it is also a structural element it may be very cost effective. The homes in Village Homes almost all have concrete slab floors because they are both the most economical floor for use here and very effective for thermal storage if they are not carpeted. Homes in the development almost all have tile floors throughout most of the house for good storage. Painted concrete floors have also been used. And some houses have parquet floors which are acceptable if the wood is not too thick.

These basic concepts and designs are used to build the energy-efficient and comfortable houses in Village Homes. The following chapters describe them all in greater detail. They have been divided in eight categories, primarily to make them easier to discuss, but in reality they all fall along one contin-uum and are arranged in order, from the simplest to the most complex. For each type a thorough and detailed discussion of the basic design is followed by a more cursory look at the houses of that type in Village Homes.

WATER STORAGE CONTAINERS

Type	Source	Size	Volume	Raw Cost/Gal.	Installed Cost	Notes
Tank	local welder	any, but a 1½′ × 3′ × 6′ module is recommended	any	30¢-$1.	50¢-$1.20	Aesthetic, easy to install, effective.
Drums	drum mfg., chemical supply, etc.		30 or 55 gal.	10-45¢	20-70¢	Cheap, readily available, hard to clean, must be stacked carefully.
Culverts	pipe supply, scrap yards	12″ + diameter	depends on length, diameter	30-50¢	50¢-$1.	Strong, attractive to some. Good where floor space is tight. Make sure installation is seismic safe.
Glass bottles	various	varies	up to 10 gal.		free to $2./gal.	Cheap, readily available, must seal carefully. Installation difficult, seismic problems.
Metal cans	container co., restaurants	varies	to 5 gal.	to 60¢ per gal. for 5-gal. can	65¢ for 5 gal. cans	Good for furniture modules. Slow to install. Would need bracing for tall stacks.
Plastic bottles and drums	various: grocery stores, etc.	varies	commonly 1-5 gal.	10-50¢	15¢-$1.	Cheap, may provide pleasant light, no rust. Slow to install. Installations may be weak.

Chapter Three

The Solar Homes of Village Homes

In this chapter the various types of solar homes in Village Homes are described in greater detail and several houses of each type are illustrated and described. Each type is introduced with an overview describing the basics of design, performance and cost, operation, maintenance, and transferability of the system to other climates.

It is difficult to compare the performance of these different houses because performance depends in considerable measure on the users. Thus a house that might provide 90% solar heating for a family that operates it conscientiously and allows the temperature to vary between 60-80°F may only give 60% when poorly operated over a narrower temperature range. Descriptions of performance are further limited by the lack of performance data in Village Homes. Only three houses have been monitored (Maeda, Starr, Bainbridge) and only two of these have been monitored extensively (Starr, Maeda). And these were operated differently, so results are not easy to compare.

Therefore, the estimates included should be taken for what they are: a rough estimate of performance for an average user. They are based on owners' descriptions of comfort, utility bills, and evaluations of basic design. They should be accurate within 10-20%.

The percent solar heating (percent of house heating provided by the sun) includes internal heat gain as part of the solar system because this gives a better view of the performance of an energy-conserving passive house. If it is excluded, the more efficient the house becomes, the lower the solar fraction drops. Suppose a house with 60% solar, 20% internal heat gain, and 20% backup heating is made twice as energy conserving with better insulation and weather stripping. The internal heat gain stays the same but may now provide 37% of the demand. Since this heat is uniformly distributed, the backup heater will still be required, but less frequently, meeting say 13% of the demand. As a result, the solar fraction will drop to 50%. Yet the house is unquestionably better and uses considerably less auxiliary energy for heating. Counting internal heat gain as part of the passive solar gain gives a better picture of the house, as this house would improve from 80-87% solar. In the house descriptions that follow, percentage solar heating includes normal internal gain.

Price is equally difficult to compare and a price range rather than exact price is used. The same house built by different builders, and in different areas, may vary by 20% or more. The use of a price range allows for fairly good cost comparison. Price ranges represent construction cost, excluding the lot, in 1978. Costs in construction have increased at about 1½% a month, and a house which

A unique housing development incorporates a natural drainage system, native plantings and extensive bikeways. All lots orient south to maximize solar access.

was in the $20-35,000 price range in 1977 could well be in the $40-50,000 price range in 1979. House costs in Village Homes have been very competitive with nonsolar houses. With the passage of California's 55% Tax Credit in 1977, cost of simple passive houses actually dropped to as much as $800 below the cost of a comparable nonsolar house.

The order chosen for presenting the different house types is based on their system, simplicity, cost, and performance. Like all classifications it is to a certain extent arbitrary but in general the complexity and cost increase through the chapter. The classifications include: the simple passive house, the water wall house, the clerestory house, the solar greenhouse, the skylight house, the suncatcher house, houses that use an active system to back up a passive system, and finally the houses that are primarily active with passive features.

	graphic symbols
— — · · — —	property line
▦	thermal mass
▦	floor-hard surface
▭	floor-soft surface
▨	concrete or paving
◼	fireplace(wood stove)
▬	wall
▧	insulation
⊐▮⊐	bicycle
✳ ◌	landscape
▲	entry
⟿	air flow
⌇⟶	heat radiation
⊠	fan

Simple Passive

Basics

The simple passive house combines all or almost all of the features described in the previous chapter on basic concepts and details to provide a comfortable and energy-efficient house much of the year. We have called it the simple passive house because it relies on good design and standard construction rather than any complex or expensive solar hardware.

The simple passive type house is the most commonly used in Village Homes because it is simple, economical, and looks like a standard house both inside and out. The solar orientation and energy conserving shell reduces the size of needed backup heating and eliminates the need for air conditioning equipment, allowing the total price to be kept in line with standard housing.

This house, however, does have two weaknesses: it doesn't heat the north side of the house very well and it may overheat on sunny winter afternoons.

Performance and Cost

The most important factor in design of the simple passive house is the south orientation of the building. Most of the windows are placed on the south and are shaded by an overhang or arbor during the summer and are exposed to the winter sun. This simple step ensures that the house will be able to use the winter sun for heating. The energy collected by the south windows is stored in the tile-covered concrete slab and internal mass of walls and ceiling.

Operable windows to the north and south allow for good cross-ventilation on summer nights. This allows the cool sea breezes to rapidly cool off the house once they reach Davis. The interior mass can store this coolth for use during the day and can make living without an air conditioner possible, although temperatures may creep into the low to mid-80s inside on some very hot summer afternoons and may stay there into the evening when the sea breezes don't blow. Energy use for cooling is typically

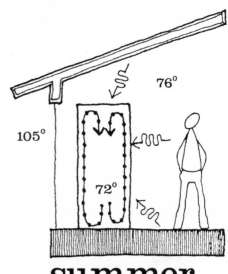

summer

limited to one or two fans rather than an air conditioner, with energy use for cooling less than 1/10 that for a standard nonsolar house in the area with air conditioning.

During the winter, backup heating is usually needed only at night and on cloudy days. Fortunately the coldest days are those following the passage of storm fronts and are usually very clear and sunny with the sun providing all or almost all of the needed heating during the day. The typical heating bill is about one-half that of a standard nonsolar house.

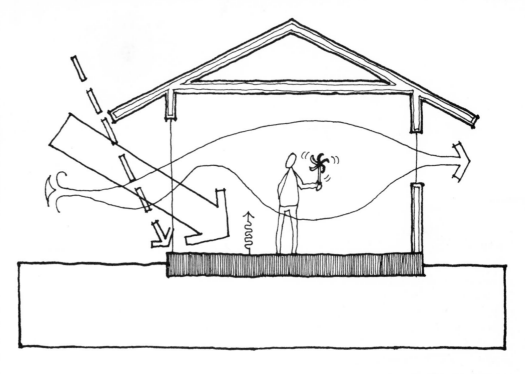

Operation

Operation couldn't be much easier. Drapes are opened in the morning on sunny winter days and closed in the evening. On summer days drapes are usually closed in the morning and opened at night; windows are opened in the evening when the sea breezes arrive and are closed in the morning when it starts to warm up.

Operation isn't foolproof, however; on a winter day everyone might leave the house in the morning while it is foggy outside. They must decide whether to open the curtains in the hope the sun will come out or leave them closed and miss the sun if it does come out. Partly cloudy days create similar problems.

Maintenance

Drapes and shutters must be kept in order. Windows should be cleaned occasionally, and the backup system, either gas or wood, should be checked every year. Gas furnaces should be cleaned and tuned periodically to burn efficiently. The backup wood heater, most commonly a Franklin-type stove, should be cleaned and cracks resealed if necessary. The chimney may need to be cleaned as well.

Transferability

The simple passive house will work well almost anywhere in the United States. It reduces unwanted heat gain in the summer and it provides considerable solar heating in the winter. This is a desirable feature even in most of the desert areas where the winter comfort zone may be 10-15°F above that in the cooler Davis climate. The placement and type of windows on the east and west should be dictated by local wind patterns to ensure good cross-ventilation. These may need exterior shades in many areas of the United States in the summer.

The Rob Thayer Residence

Design: *Mike Corbett*
Construction: *Mike Corbett*
Price Range: *(Adj. 1978)*
$34–38,000
Living Area: *890 sq. ft.*
Built: *1975*

South view with brick courtyard and living trellis.

Rob Thayer both teaches landscape architecture at the University of California in Davis and has a private landscape practice. His east side two bedroom, one bath, commonwall house is entered through a courtyard. He has converted one bedroom into an office. His house and the common area behind it have benefitted from his skill in landscape design and are among the most attractive in Village Homes. His living room has a built-in planter in front of the south window and often threatens the whole room with its lush vegetation.

The System

This house relies on good orientation and window placement for solar heating and natural cooling. Thermal storage is provided by tile floors in en-

Lush interior plantings behind south windows.

42

try, dining, kitchen, and living rooms. One water-filled culvert was added to the south bedroom in 1978 for additional thermal storage. A wood stove and gas furnace are used for backup. Cross-ventilation, shading, and light exteriors are used for natural cooling.

Performance

The house gets a bit warm on a few summer afternoons and evenings, but it can still be called full cooling. The solar contribution to heating is probably about 50-60%. Wood is used for some of the backup heating. Utility bills average about $10 a month in the summer and $20 a month in the winter.

Improvements

Rob would like to install a security ventilation system with a through attic fan to pull cool air in through the house and out the attic. He has also considered adding thermal drapes and shutters and adding more thermal mass to further increase performance. However, the best spot for the mass would be where the planters sit, and the plants have won so far. The courtyard fence blocks some of the low winter sun.

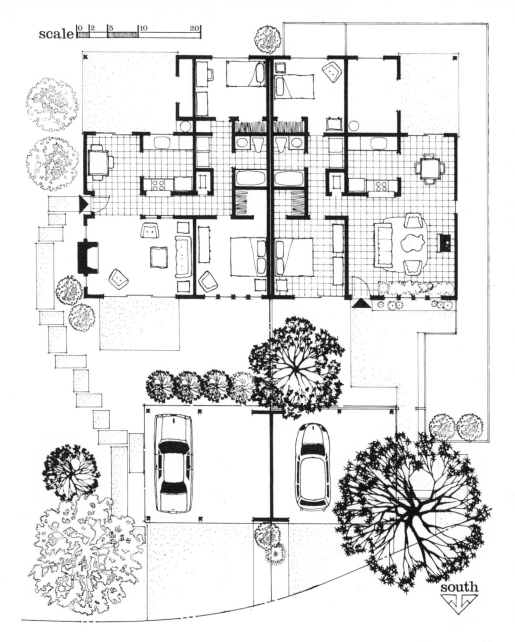

scale 0 2 5 10 20

south

43

The Longshore Residence

Design: *John Hofacre*
Construction: *Mike Corbett*
Price Range: *(Adj. 1978)*
$57–61,000
Living Area: *1,470 sq. ft.*
Built: *1976*

Betty Longshore's house is a two-bedroom, one-bath home with an open plan and large rooms, designed especially for its owner. Betty refuses to use the city's garbage service and recycles, burns, or composts all of her wastes. She has no garbage disposal or dishwasher. An open courtyard has been added to the street-side of the house.

The System

The Longshore house is a simple passive house with a bank of long, narrow windows on the south for heat collection and a tiled window seat and exposed tile floor throughout the house for storage. Blinds are used on the south windows for sun control and provide indirect lighting. The walls use 2 × 6 framing with R-19 fiberglass insulation.

Arbors and overhangs provide summer shade for the windows and south wall. Betty grows a productive wall of beans across her south windows in summer. The house also has very good cross-ventilation for cooling. Backup heating is provided by two Jøtul airtight stoves, and a forced air gas furnace.

Performance

The extra shade from the landscaping helps keep the house cool and comfortable in the summer. The arbors and plants die off in the fall and are carefully trimmed and cleaned up to provide the best possible exposure in the winter. The solar system meets about 50% of the winter heating demand and the wood stoves provide most of the backup heating. Utility bills for two people average about $10 in the summer and $20 in the winter.

Improvements

The solar performance of this home could only be improved by moving it

Rustic wood trim accents south windows.

out of the simple passive category with the addition of water mass. Adding insulated drapes or shutters to the windows would also help. Perimeter slab insulation should have been used.

Tiled window seat and floor absorb heat, and thin venetian blinds provide shade in summer.

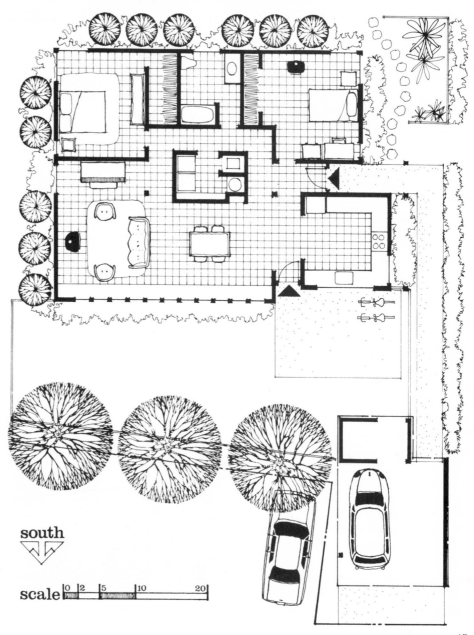

south

scale 0 2   5   10   20

The Hagerty Residence

Design: *John Hofacre*
Construction: *Mike Corbett*
Price Range: *(Adj. 1978)*
$46–50,000
Living Area: *1,382 sq. ft.*
Built: *1976*

An open living area provides a spacious feeling.

Don and Rebecca Hagerty came to Village Homes fairly early and their satisfaction with their three-bedroom, two-bath house is now based on considerable experience. The kitchen, dining, and living rooms are all open and have a southern view onto the greenbelt. A west-facing sliding glass door provides access through the dining room to a patio. The Hagertys find the natural cooling particularly appealing. Their garden is lush and provides harvests much of the year.

The System

The Hagerty house relies on simple passive design for both heating and cooling. Tile floors in the entry, dining room, and kitchen and a masonry fireplace help store heat in winter and coolth in summer. Three eight-foot-wide

Three pairs of sliding glass doors take the place of windows on the south.

sliding glass doors on the south collect winter sun for heating and provide natural ventilation for summer cooling. South windows are shaded in summer and light exterior colors help keep the house comfortable. Backup heating is provided by a central fireplace with heatilator and a gas furnace.

Performance

The house provides full cooling and about 50-60% of the winter heating. The lack of solar hot water, and some inefficient appliances, put their total bills for two people to a January high of $22 and a July low of $17 and their monthly gas bills between $11 and $4.

Improvements

The owners would like to add a security ventilation system that will give them better cooling without fully opening doors and windows. The performance of the house would also benefit from insulated drapes or shutters and additional thermal mass. The patio sliding glass door to the west is a source of unwanted summer heat gain. The Hagertys are now using a rollup shade on it but plan to build a wood patio cover. An airtight, wood-burning stove would be far more effective in heating the house than the masonry fireplace.

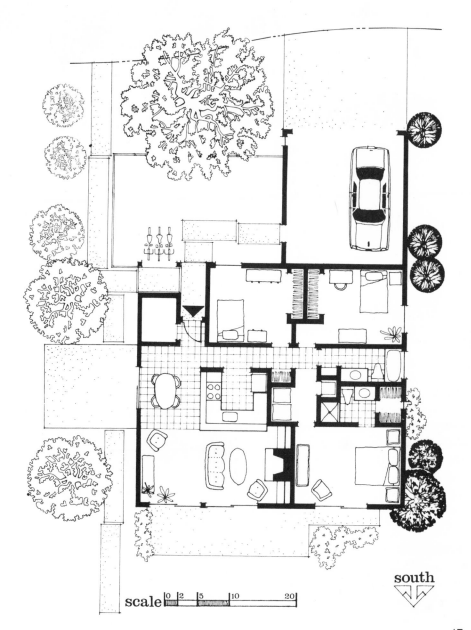

scale 0 2 5 10 20

south

The Vasak Residence

Design: *John Hofacre*
Construction: *Mike Corbett*
Price Range: *(Adj. 1978)*
$44–48,000
Living Area: *1,152 sq. ft.*
Built: *1976*

Lacey Vasak has a large loft, which has come in handy as a bedroom for overnight guests, in her two-bedroom, one-bath passive solar house. Window seats add a special touch to the bedrooms. Stained glass windows next to the front door and in the loft add a pleasing glow to the interior spaces. The living room is decorated with a brick fireplace that includes a heatilator for greater heating efficiency. Lacey's private, street-side courtyard is fenced by a vine-covered stucco wall. A deck and orchard are included on the north side of the home.

The System

The Vasak house is a classic example of simple passive design. Orientation is to the south for winter solar heating and summer cooling breezes. Exterior wood shades and grape arbors provide almost full shade in summer. The shading, light exterior colors, and cross-ventilation provide very good cooling. Storage for winter heating and summer cooling is provided by tile floors in the entry, dining room, and kitchen.

Performance

This simple passive system provides about 50% of the heating. The natural cooling works very well because there are few windows on the east and west walls. The loft stays warm even on the coldest days and allows Lacey to stay comfortable without turning on the backup heater. Even without solar hot water, the utility bills for the house ranged from a high of $14 to a low of $11 and the gas bills were between $6 and $3.

Improvements

Obviously the house works well but it could be improved. A solar hot water heater could be added; more mass and insulated drapes and shutters would improve heating and cooling performance; and an airtight wood stove would perform more efficiently than the masonry fireplace.

The sleeping loft above living room is reached by a ladder.

Grapes shade the south glass in the summer.

A stucco fence with wood accent creates a private south-side courtyard.

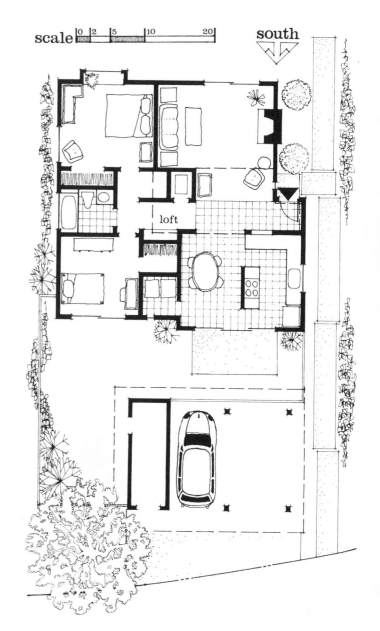

scale

south

loft

49

The Inouye Residence

Design: *Mike Corbett*
Construction: *Mike Corbett*
Price Range: *(Adj. 1978)*
$56–60,000
Living Area: *1,600 sq. ft.*
Built: *1977*

Sliding glass doors provide more glass area at less cost than windows.

Patsy and Tim Inouye are very pleased with the open plan of their two-bedroom, two-bath home with a study loft. A staircase winds its way up to the second floor and provides an attractive architectural detail in the living room. A courtyard makes a pleasant entry and the deck on the south side helps expand the living space.

The System

The Inouye house is oriented for good solar gain through the south windows in the winter with tile floors in the entry, dining room, and kitchen, and a masonry fireplace for storage. Backup heating demands are met by a fireplace with heatilator and a gas furnace. Cooling is provided by careful shading, use of light exterior colors, and good cross-ventilation. It is augmented with a portable fan on the hottest days. Hot water is supplied by a thermosiphon solar water heater with a 48-square-foot panel and an 82-gallon tank.

Performance

The house provides full cooling and 50–60% of the heating. The heatilator fireplace is used as the primary backup heater but isn't needed even on cold days if the sun is shining.

Last year the backup heating required only a half cord of wood. The total utility bills for two people range from a January high of $22 to an August low of $15, and gas bills range between $1.50 and $11.

Staircase broken with landing adds interest to living room.

Improvements

The house would work even better if it had additional mass and insulated shutters and drapes. An airtight stove would provide more efficient heating.

50

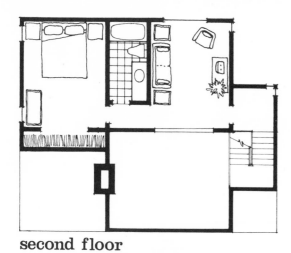

second floor

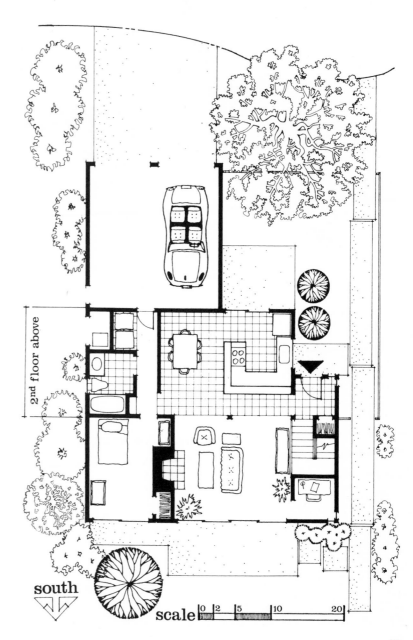

The entry is through the north courtyard.

2nd floor above

south

scale 0 2 5 10 20

The Ferris Residence

Design: *John Hofacre*
Construction: *Mike Corbett*
Price Range: *(Adj. 1978)*
$48–52,000
Living Area: *1,550 sq. ft.*
Built: *1977*

The south elevation provides plenty of glass.

Jay and Debby Ferris are very happy with their simple passive house. They particularly like the floor plan and they use the different zones very effectively. The dining and living areas are on the south and become warm early in the morning. On cold, cloudy days a fire in the centrally located wood stove quickly warms up the breakfast area. (Open centrally located wood stoves should only be used where danger of accidental burns is minimal.) They like the bedrooms cool, and on the north side they stay comfortably cool all summer. The loft is also popular, particularly on cold winter days when it is always cozy. The home has three bedrooms and two bathrooms. The large south-facing front yard has been intensively developed by Debby and Jay with a brick patio, lawn, and vegetable garden.

The System

The house is a very simple passive house. It uses south windows to collect the sun's energy and both tile floors in the entry, dining room, and kitchen, and the interior mass of the building for storage. Backup heating is provided by a wood stove and gas furnace. Natural cooling is provided by cross-ventilation. An operable skylight near the peak provides additional ventilation. Cooling load is reduced by south overhangs, light exterior colors, and minimum east and west glass. Water heating is provided by a thermosiphon solar system with gas backup.

Performance

The house provides full cooling, with a few warm afternoons and evenings, and about 60% of the heating. It shows very clearly what can be done simply by orienting a house properly and designing it well.

Improvements

This house would work much better if it had additional thermal mass. One of the nice things about these simple passive houses is the ease with which mass could be added and presumably will be added when utility prices increase. The house would also work better if insulated drapes or shutters were used throughout.

Wood-burning stove is centrally located for good heat distribution.

A loft overlooks the dining room.

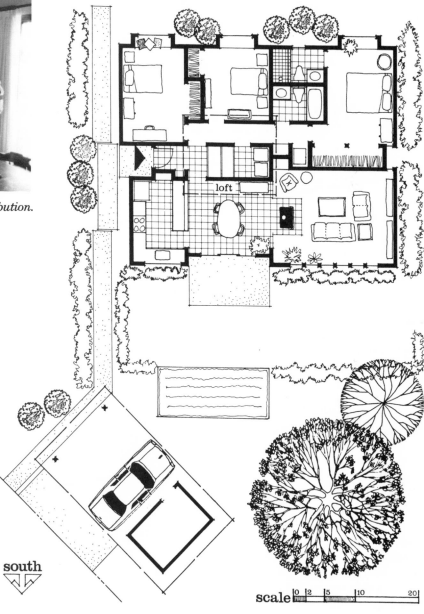

loft

south

scale 0 2 5 10 20

The Dewees Residence

Design: *John Hofacre*
Construction: *Mike Corbett*
Price Range: *(Adj. 1978)*
$48–52,000
Living Area: *1,570 sq. ft.*
Built: *1978*

by cross-ventilation, arbors with grapes, south overhangs, light exterior colors, and minimal use of glass on the east and west. Backup heating is provided by a wood stove centrally located in the dining room and water is heated with a thermosiphon solar system with a gas backup.

Performance

The house provides full cooling with only a few warm afternoons and evenings. The solar system provides

Kitchen island provides handy counter and storage space.

The Dewees family worked very closely with the designer to design their three-bedroom, two-bath home with a large family kitchen. The attic space of 435 sq. ft. was left unfinished to reduce initial cost. The living and dining areas are open and accented by a staircase to the attic and sliding glass doors which open to a south deck. A trellis over the deck will be covered by deciduous vines in the future so that summer sun won't reach the south glass.

The System

The Dewees house is a simple passive house and uses the south windows to collect the winter sun. Storage is provided by slab with tile covering in the entry, dining room, and kitchen, and interior mass of the walls. Cooling is aided

Wood deck enhances glass on southern exposure.

54

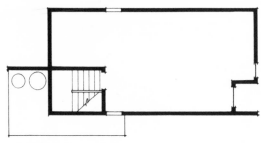

second floor

about 60% of the winter heating and wood meets much of the other 40%. The solar system provides all of the hot water in the summer and much of the hot water in the winter.

Improvements

The addition of added thermal mass in the house would be desirable. Thermal drapes and shutters would also help improve performance. Sliding glass doors have been fitted with locks which allow the door to be secured while several inches open. A fan could be installed to increase the circulation of night air into the house.

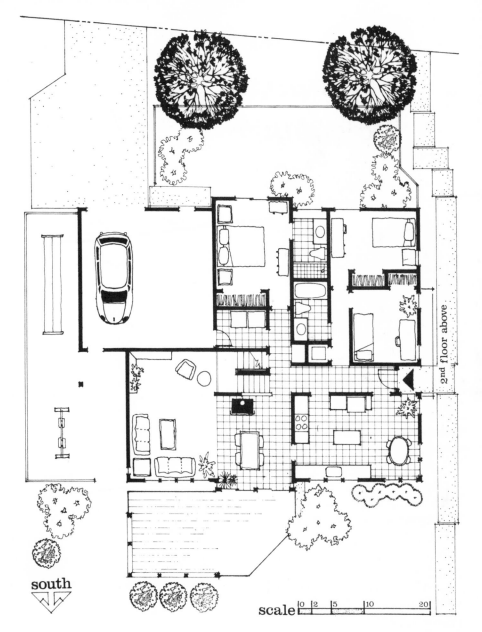

2nd floor above.

south

scale 0 2 5 10 20

The Auman Residence

Design: *John Hofacre*
Construction: *Mike Corbett*
Price Range: *(Adj. 1978)*
$50–54,000
Living Area: *1,615 sq. ft.*
Built: *1978*

Expansive glass on south provides light and heat to living spaces.

Michael and Shirley Auman wanted play space for their young son and daughter. They got it in this attractive three-bedroom, two-bath, two-story home. The two small upstairs bedrooms open onto a spacious loft, used as a playroom.

The System

The Auman house uses the 220 sq. ft. of south windows and tile-covered concrete to collect and store the winter sun's heat. A Frontier airtight stove and gas furnace are used for backup heating. Cross-ventilation and solar control provide natural cooling. A wind-powered vent is located at the ceiling peak to assist cross-ventilation. A thermosiphon system with gas backup is used for domestic water heating.

Performance

The house is fully cooled naturally. The passive solar features should provide about 60% of the heating demand with wood making up much of the rest. The thermosiphon should meet 70% of the hot water demand over the year.

Improvements

Thermal mass and insulated drapes and shutters would be desirable. An arbor with grape vines to the south is planned and will also help in cooling. In designing houses with "open plans" for natural heat distribution, attention should be paid to sound control.

Loft overlooks the living room.

second floor

Kitchen is set off from dining area by high-backed counter.

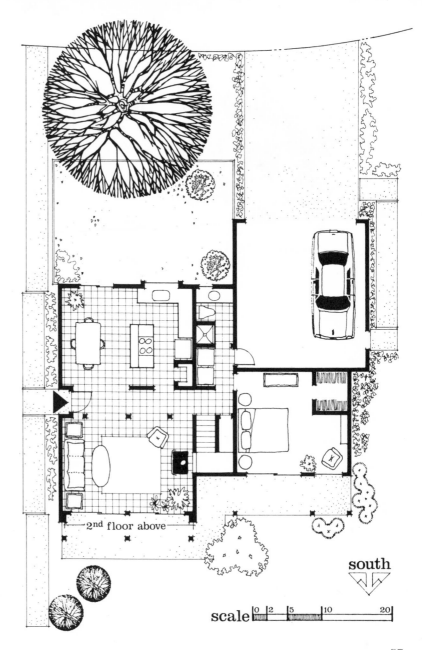

2nd floor above

south

scale 0 2 5 10 20

The Perkins/ Riviello Residence

Design: *Mike Corbett*
Construction: *Mike Corbett*
Price Range: *(Adj. 1978)*
$38-42,000
Living Area: *1,030 sq. ft.*
Built: *1978*

South windows are maximized, west windows used sparingly.

Although it is small, the open plan and the many south windows make this house feel spacious. Tony Perkins and Mark Riviello, the part-time college students who live here, are very pleased with this two-bedroom, one-bath house. They particularly like the views from the kitchen, dining and living areas, provided by the double-pane sliding glass doors.

The System

This simple passive house uses the 150 sq. ft. of south sliding glass doors to collect the winter sun for heating. Storage is provided by tile and the slab in the entry, kitchen, dining room, and part of the living room. A Frontier airtight stove and gas furnace are used for backup heating. Cooling utilizes the cool breezes, good solar control with light exterior colors, few windows on the east or west, and overhangs on the south. A thermosiphon solar water heater with gas backup provides the hot water.

Performance

The Perkins/Riviello house will get about 60% of its heating from the sun. The thermosiphon solar water heater should provide about 70–80% of the hot water. And the passive features will provide all of the cooling.

Improvements

The overhangs on the south are not wide enough and an additional wood strip or grape-covered trellis should be added. Insulated shutters and drapes would also help. The large amount of south glass warrants more internal thermal mass and could bring the heating potential up to 80%.

second floor

Kitchen is recessed to accommodate more south glass.

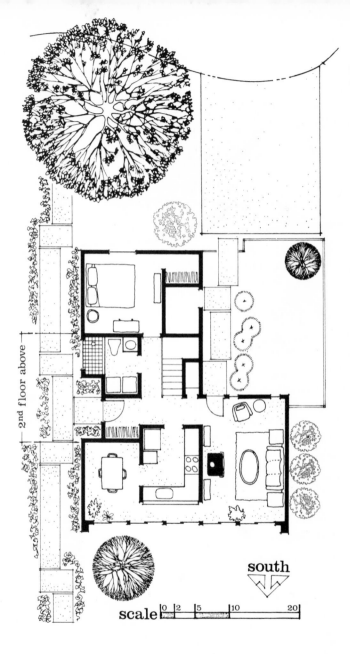

2nd floor above

south

scale   0   2   5   10   20

The Cech Residence

Design: *John Hofacre*
Construction: *Mike Corbett*
Price Range: *(Adj. 1978)*
$48–52,000
Living Area: *1,564 sq. ft.*
Built: *1977*

Joe and Mary Cech and their two young boys wanted their kitchen, dining and living areas to be open to allow for casual entertaining in their three-bedroom, two-bath house with study. The adult and children's bedrooms are located on opposite sides of the house to provide quiet and privacy for the parents. A private courtyard on the street side and a lawn on the greenbelt provide different outdoor living areas to suit different uses.

The System

The Cech house relies on building shape and form for views, solar heating, and natural cooling. Furniture placement in the rooms dictated windows about 30 in. from the floor so there is not as much south glass as there could

have been, but it still includes 144 sq. ft. The overhangs were reduced to 2 ft. 6 in. to allow greater sun penetration during winter. Thermal storage is provided by the concrete slab with tile floors in the entry, dining room, hall, and kitchen. A Jøtul airtight wood stove and gas furnace provide backup heating. Full natural cooling is realized by cross-ventilation and careful solar control with overhangs, light exterior colors, and landscaping. Hot water is heated with a thermosiphon solar water heater and a gas backup.

Performance

The Cech house performs well for both heating and cooling. Temperatures usually don't rise above 80°F in the summer and rarely drop to 60°F in the winter. Utility bills have averaged $14 per month.

Improvements

This house could use more south glass. Either water tanks or culverts would be easy to retrofit. Insulated drapes and shutters would be desirable.

Picture windows accent south elevation.

Eating bar backed up to kitchen counter makes efficient use of space.

Centrally located fireplace on raised tiles.

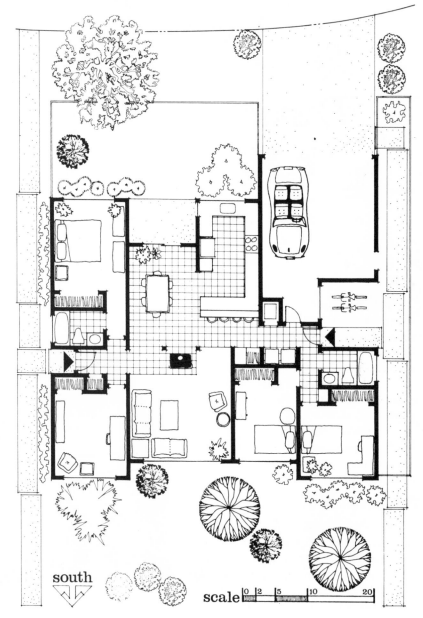

south

scale 0 2 5 10 20

The Polito Residence

Design: *Mike Corbett*
Construction: *Mike Corbett*
Price Range: *(Adj. 1978)*
$40–44,000
Living Area: *1,208 sq. ft.*
Built: *1977*

They use the attic access door to provide added ventilation in the summer. A central Frontier airtight wood stove provides all the backup heating, although a gas furnace was installed. A thermosiphon system with gas backup handles domestic hot water demand.

Performance

The Polito house is one of the best simple passive homes. Natural cooling is very good and indoor temperatures stay below 80°F if the house is opened at

Rug adds comfort but reduces tiles' thermal storage advantage.

Vito and Esther Polito and their two small children have enjoyed their compact, three-bedroom, one-bath home. Both living room and master bedroom open onto the attractively landscaped courtyard. The greenbelt side of the home is terraced with vegetables and ornamental plantings. The Politos make efficient use of their home's solar capacity, opening and closing drapes and windows at the appropriate times.

The System

The Polito house relies on good orientation for heating and cooling. Thermal storage is provided by tile floor slab in the entry, dining room, and kitchen. Cross-ventilation is good with several large windows on the north and south.

Patio doors open to a cozy private courtyard on the south.

Centrally located wood stove provides sufficient backup heating for the entire house.

night. The solar system provides about 60% of the needed heating. Utility bills for four range from an August low of $24 to a January high of $28. The gas bills range between $3 and $6.

Improvements

The Politos would use tile throughout the house if they were to build again. Water-filled tanks or culverts would be even more effective. Thermal drapes or shutters would also help improve the performance of the house. The Politos would also add a system for security ventilation.

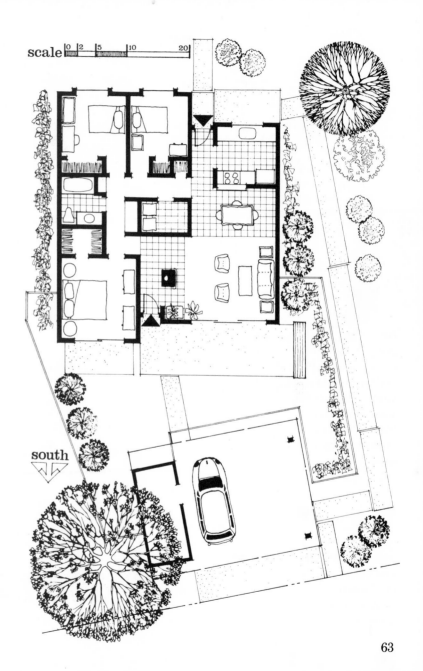

scale 0 2 5 10 20

south

63

The Wisner Residence

Design: *Mike Corbett*
Construction: *Mike Corbett*
Price Range: *(Adj. 1978)*
$64–68,000
Living Area: *1,750 sq. ft.*
Built: *1978*

South facade offers a pleasant view out of a corner kitchen window.

This three-bedroom, two-bath house has a living room accented by window seats. A family room looks out on an intimate courtyard formed by the wall of a detached garage. Corner windows in the eating nook and kitchen afford a pleasant view of the greenbelt. The south arbors are covered with a lush grapevine.

The System

The Wisner home uses windows on the south to collect the winter sun. A concrete slab with tile floors in the entry, dining room, family room, and kitchen provides thermal storage. A heatilator system in the fireplace and a gas furnace provide backup heating. Night ventilation provides the cooling for the house, which also has an air con-

Tiled entry and a large brick fireplace add mass to the house.

Wood trellis supports vines.

ditioner. Minimal use of east and west glass, south overhangs, arbors, and light exterior colors also help keep the house cool and comfortable. A pumped solar system with electric backup is used for hot water.

Performance

The passive design features minimize need for the air conditioner. The solar heating handles about 50% of the heating demand with wood and gas making up the rest. The solar system provides all of the hot water in summer and much of it in the winter.

Improvements

Adding insulated drapes and shutters would help both cooling and heating performance, but the biggest improvements would involve moving the house from simple passive into the water wall classification. This added mass is really needed for better performance.

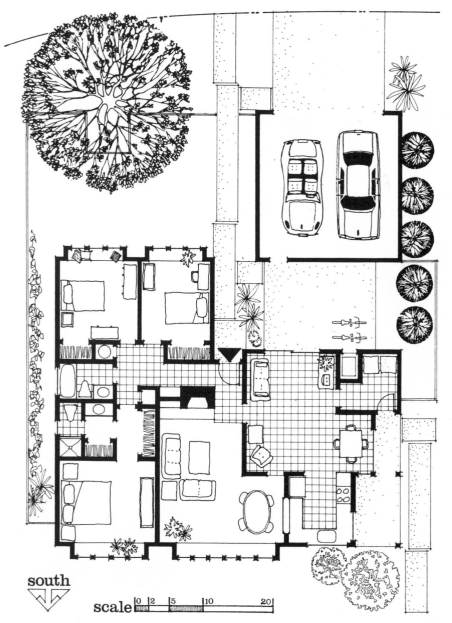

south

scale |0 |2 |5 |10 |20

The Rapaport-Fereres Residence

Design: *Mike Corbett and John Hofacre*
Construction: *Mike Corbett*
Price Range: *(Adj. 1978) $54–58,000*
Living Area: *1,558 sq. ft.*
Built: *1978*

This three-bedroom, two-bath, two-story house is enclosed on the south side by a courtyard. Special design features include a wood bookcase built into the upstairs hallway and northeast and northwest corner windows capturing a greenbelt view.

The System

This simple passive house relies on south windows for solar collection and tile floors on slab in the entry, dining room, and kitchen for storage. Cooling is provided by cross-ventilation at night. Central air conditioning and heating are used for primary backup although a fireplace with heatilator is also used.

Performance

This house has met about 90% of the cooling demand and about 50% of the heating demand naturally. The air conditioner is only used a few days each year and the furnace is used primarily on cloudy or foggy days in the winter. The heatilator seems to work well only with a large fire.

Improvements

The addition of thermal mass for storage in the house and insulated shutters or curtains would be very helpful. The owners would also make some minor changes in the house plan. These include adding more storage, more roof overhang, and a more sheltered entry. They would also like to be able to close off the kitchen.

Second floor balcony leads to bedrooms.

The north side features corner-view windows.

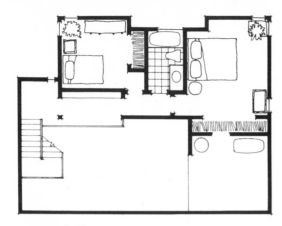

second floor

High ceiling makes living room feel more spacious.

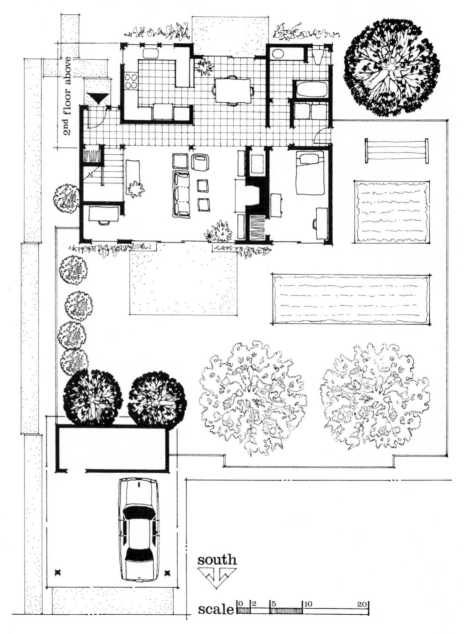

2nd floor above

south

scale 0 2 5 10 20

The Diaz Residence

Design: *John Hofacre*
Construction: *Mike Corbett*
Price Range: *(Adj. 1978)*
$62–66,000
Living Area: *1,830 sq. ft.*
Built: *1977*

Sandra and David Diaz and their three teenaged children are pleased with the low energy bills and the comfort of their four-bedroom, two-bath home. The high living room windows are much appreciated for the view and natural lighting. A formal courtyard is located on the front side of the house and a large rooftop deck has proven to be a very popular place for the family to spend their summer evenings.

The System

Extensive south-facing double-pane sliding glass doors (170 sq. ft.) on the Diaz house collect the sun for storage in the tile-covered slab and the interior mass of the building. Backup heating is provided by a gas furnace, and a wood stove which is located between the living and kitchen/nook areas. Natural cooling is supplied by cross-ventilation and careful solar control. Exterior surfaces are light in color; windows are shaded with overhangs; and exterior surfaces are shaded by grape arbors to help reduce cooling loads.

A horizontal tank thermosiphon solar system (82 gallons) is used for heating hot water. A 40-gallon gas backup heater is also included.

South glass is protected from strong summer sun by arbor.

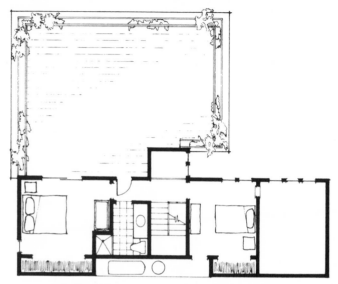

second floor

Large roof deck invites sunny outdoor relaxing.

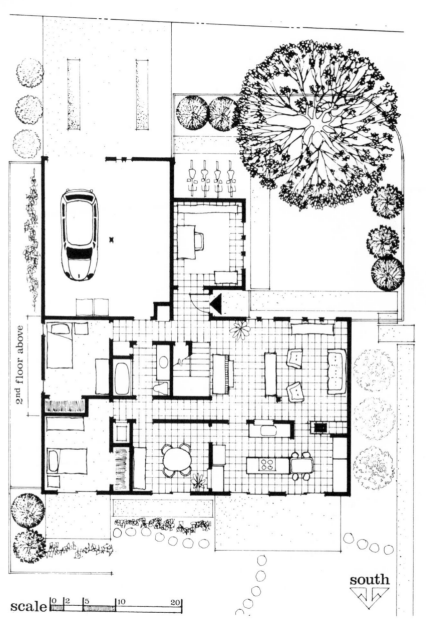

2nd floor above

south

scale 0 2 5 10 20

Performance

The house has been very comfortable and economical. With the thermostat set at 62°F the furnace very rarely turns on. The solar system meets most of the hot water demand for the year. No backup cooling is needed. The roof deck is used for sleeping on the hottest summer nights.

Improvements

The Diaz house would work even better if it had added internal mass and insulated shutters and drapes. An automatic security ventilation system and less north glass and more south glass would also be desirable. Because the bottom floor is completely tiled, the house is a little noisy. Acoustical ceilings, textured drapery fabrics, wall hangings, and soft furniture would reduce the problem.

South glass doors make dining room seem bigger than it is.

Wood stove separates living room from dining area.

70

The north side features a courtyard and gracious roof deck.

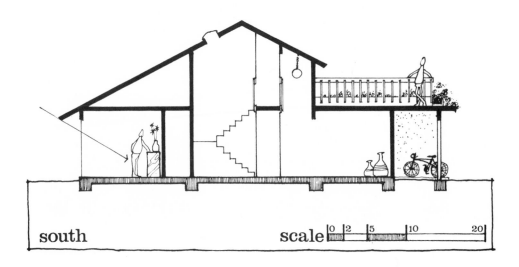

south   scale |0 |2 |5 |10   20|

The Squier Residence

Design: *Mike Corbett*
Construction: *Mike Corbett*
Price Range: *(Adj. 1978)*
$40–44,000
Living Area: *1,120 sq. ft.*
Built: *1977*

An attractive stained glass panel is lit by sunlight.

Vicki Squier helped Mike Corbett design her two-bedroom, one-bath home. She has been very pleased with the house, community, and the quality of her investment. The loft in the house is a very cozy spot, particularly on cold winter days and is sometimes used for sleeping. Vicki has built a comfortable space for outdoor living with the addition of a south-side patio and trellis. She maintains an extensive vegetable garden in her greenbelt area.

The System

This house uses a bank of sliding glass doors on the south side of the house as the solar collector. Thermal storage is provided by the concrete slab that is tiled in the entry, dining room and kitchen, and the interior mass of the building. A brick fireplace and gas heater provide backup heating. Natural cooling uses cross-ventilation at night, overhangs to shade south glass, elimination of east and west windows, and use of light exterior colors. A thermosiphon solar water heater with an 80-gallon storage tank and gas backup is also included.

Performance

The house provides full cooling with a few warm afternoons and evenings in July and August when a portable fan is used. Vicki has aided the cooling capacity of her home by growing grape vines on a trellis across the south side of the house. She also built a patio cover with removable shade panels to improve summer cooling. The solar heating is also reasonably good, and the system provides around half of the heating demand. The water heater provides full heating in the summer and much of the heating in the winter.

Improvements

This house is a classic simple passive house and could be improved primarily by adding thermal shutters or curtains and thermal mass. Full-grown trees to shade the roof will help with summer cooling. The cost of a complete retrofit would be about $1,200 for a system that would provide 70–80% heating and better cooling in the summer. However, the added mass would usurp some of the floor area. Security ventilation would also be a welcome addition.

A grape-covered arbor shades south glass only in summer.

Southern exposure adds warmth to the dining room.

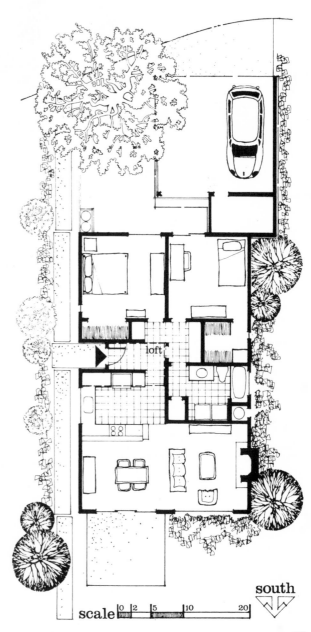

loft

The Mackey Residence

Design: *John Hofacre*
Construction: *Mike Corbett*
Price Range: *(Adj. 1978)*
$37–41,000
Living Area: *945 sq. ft.*
Built: *1976*

Betty Mackey lives in this compact, two-bedroom, one-bath house with her sister. The home was designed to fit on the very narrow lot and has an open wood deck to the south and a fenced area to the north. Three long, narrow windows brighten the hallway to the bedrooms.

The System

The Mackey house relies on south windows to collect winter sun for heating and the tile in the entry, dining room, kitchen, and hallway, and interior mass for heat storage. No windows on the east and west and light exterior wall colors help keep it cool, along with an air conditioner. Backup heating is provided by gas.

Performance

The limited storage in the floor and interiors keep the solar heating at about 50%. The house is built for natural cooling, but because of security concerns, the windows are not opened at night and an air conditioner is used instead. With the air conditioner the utility bills for two people hit an August high of $34 and a January low of $26. Gas bills ranged from $5 to $15.

Improvements

The house would perform better with insulated drapes and shutters and more internal mass. It would also do better if the thermal efficiency of the shell were improved with increased insulation and better weather stripping. The exterior masonry fireplace could be replaced with an airtight wood-burning stove. Less north window glass would have been preferable. The addition of security ventilation is necessary.

Protected side entry featuring narrow vertical window.

Narrow south windows accent the hallway.

The house faces south common area. Note raised garden beds.

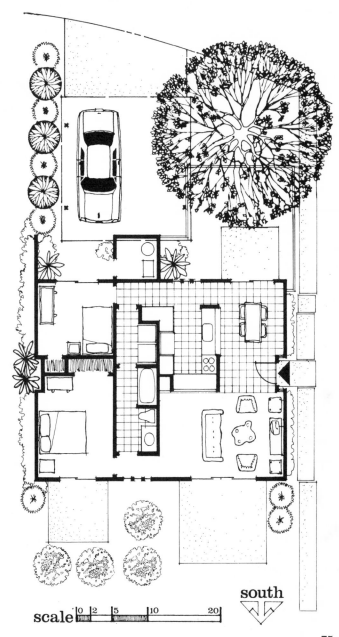

scale |0 |2 |5 |10 20

south

Water Wall

Basics

The water wall houses are probably the most cost effective solar houses in Village Homes. These houses can be described as simple passive solar homes that have large water masses in front of the south windows to prevent overheating on spring days, store the sun's energy for use on winter nights, and store coolth from the sea breeze for daytime cooling in the summer. They have been called water wall houses because the amount of water storage needed for optimum performance takes up from one-quarter to one-half of the south wall.

The water wall is essentially a large thermal battery that stores heat or coolth for use as needed. Fifty pounds of water (six gallons) per square foot of south glass is desirable to prevent overheating. Even more water is desirable to provide storage for more cloudy days in winter or more nights without cooling summer breezes in the summer. The windows in front of the water wall should of course be fully shaded in the summer and should be designed to expose the black side of the water wall to the winter sun. In addition, they should be operable, at least in part, to maximize cooling when the sea breeze is blowing. Insulated drapes or shutters between the water wall and the window are desirable but not necessary in Davis.

Performance and Cost

The water wall house can be built using standard design and construction techniques with only the addition of the water containers in front of the south windows required to make it very effective for both heating and cooling. The large area of exposed surface on the water wall also helps keep the house in the comfort range even if the air temperatures rise above or drop below what would be comfortable in a house where radiant temperatures follow air temperature more closely.

The water wall house heats up much like the simple passive house. The winter sun enters the south window and strikes the dark paint on the south side

PERFORMANCE OF A BASIC WATER WALL HOUSE[20, 21]

	Solar Contribution %	
	Heating	Cooling
San Diego, Calif.	100	86
Los Angeles, Calif.	100	86
Orange County, Calif.	96	88
San Jose, Calif.	76	100
Sacramento, Calif.	65	100
Riverside, Calif.	98	96
Contra Costa, Calif.	70	100
Santa Maria, Calif.*	99	NA
Fresno, Calif.*	85	NA
Albuquerque, N.Mex.*	91	NA
Nashville, Tenn.*	68	NA
Dodge City, Kans.*	78	NA
Medford, Oreg.*	59	NA

*without internal heat gain

of the water container where it is absorbed. As the metal on the south side of the container heats up, it heats up the water along that side of the tank. The warm water rises and is replaced by cooler water, setting up circulation which further improves the efficiency of the water wall. The circulating water keeps the south side of the tank from getting too warm and minimizes radiation losses out the window. Thus, even though the heat capacity of water is only twice that of concrete, the effective storage is three to four times as great.

The cooling cycle with the water is also similar to the simple passive house, only once again it is much more effective. The cool breeze enters the window and cools off the outside of the tank and the interior of the room. The tank cools off by direct convection and also by radiating to the cooler surfaces in the room. The cool surface of the tank is a delight on a hot summer day and its low radiant temperature, typically 70-75°F, can provide comfort even if the air temperature climbs to 80 or 82°F. If insufficient water mass is included, then the mass will track air temperature more closely.

This expanded comfort range, which really is much more comfortable than a standard house through most of its range, can provide full cooling and 75-85% of the heating required in Davis. It does this with no moving parts at a cost of only $250-500 more than the simple passive home.

Water mass will be beneficial even if it is added in rooms where the sun won't shine on it. The value of the mass for heating will be only about one-tenth that in the direct sun, or one-third of mass in the same room as the sun, but will be equal or better for cooling. This mass can be tucked away under beds, stairs, or as part of the furniture.

The tank interior can be left exposed in most cases if the tank will be sealed. If water is pumped through the tank or if the water is particularly corrosive, it may be desirable to coat the inside of the tank with epoxy, rust-inhibiting paint, or tar. A sacrificial magnesium plug added to the tank will also prevent rust. Or alternatively, rust inhibitors can be added to the water. These must be chosen to fit the particular corrosives in the water. Davis water is very hard and is very noncorrosive.

The cost of storage depends in large part on the choice of container for the water and the finish. The table, labeled Water Storage Containers in Chapter 2, lists the advantages and disadvantages of the many types of containers that can be used for water wall houses along with information on availability and cost.

The type of container and the finish in large part determine the installation cost and consumer acceptability. Specially welded rectangular tanks, as

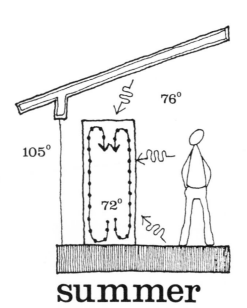

summer

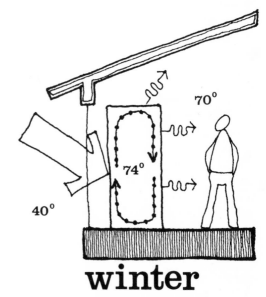

winter

used in the Bainbridge and Piña houses, appear at this time to be the most acceptable in terms of aesthetics and cleaning, and will undoubtedly be used more often in the future.

Operation and Maintenance

The operation and maintenance of the water wall house is almost identical to that of the simple passive solar home. The only additional maintenance is the somewhat awkward job of cleaning between the tank and the window. Refer to the previous section for a full description of the operation of the system. The accompanying diagrams illustrate the operation of the water wall in summer and winter.

Transferability

The water wall house is practical in most areas of the United States, although it is perhaps best suited to those areas where a hard freeze is rare (so no

antifreeze needs to be added). In colder areas the masonry wall or Trombe wall, as it is often called, may be more practical even though it costs considerably more for a given level of storage.

Some research has been done on water wall houses, and although only heating was considered in most cases, it is still worth looking at. Similar performance can be expected with most of the other solar houses if proper glass area and mass are included.

The Thigpen/ Hunt Residence

Design: *Gregory Acker, architect,*
Living Systems;
Marshall Hunt
Construction: *Virginia Thigpen*
and Marshall Hunt
Price Range: *(Adj. 1978)*
$58–62,000
Living Area: *1,600 sq. ft.*
Built: *1976*

This three-bedroom, two-bath house was the first water wall house in Village Homes. It has pleased Virginia, a contractor, and Marshall, a solar consultant, with its performance and comfort. The three bedrooms work very well as two offices and one bedroom. A comfortable loft is hidden near the peak. The structure is interesting, as the design shows off the handsome roof trusses. It is a delightful house with very attractive spaces and comfortable temperatures all year. The home is a *Sunset* magazine award winner for energy-efficient design.

The System

The south wall is almost all windows. It includes a majestic two-story window wall which is doubled-glazed, and a dining room solarium that is triple-glazed. The tall window is backed with seven water-filled culverts, each 14′ × 18″. These are spaced about six in. apart to allow interesting views of the backyard. The dining room windows use the parquet-covered slab for storage. A small fan and duct are used to carry hot air from the peak to the north bedrooms. All of the backup heating is provided by a Franklin stove which sits in front of the culvert pipes.

Natural cooling is provided by cross-ventilation of the house from the cool sea breeze. An operable skylight provides increased ventilation by utilizing the thermal stack of the two-story house. An exhaust fan installed near the peak provides added ventilation. Exterior rolldown shades on the two-story window wall and the east-facing windows of the solarium, minimal window area on the east, west, and north, and light-colored exterior walls help reduce cooling demand. The light grey concrete tile used on the roof was developed by the manufacturer for this house for the same reason. The arbor over the deck

South side hosts elegant two-story glass wall and breadbox water heater.

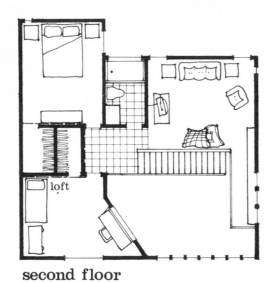

second floor

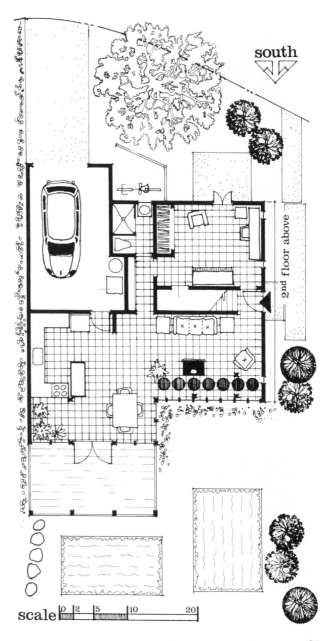

loft

Hand-operated, rolldown canvas awning provides summer shade.

south

2nd floor above

scale 0 2   5   10   20

81

Nighttime view dramatizes position of water-filled culverts.

Windows are minimized on east, north, and west sides.

Culverts rise like pillars in the living room.

has both deciduous vines and removable shade panels for good summer shade with almost full winter exposure. A casablanca fan is used for backup cooling.

Solar hot water is heated in a double breadbox system. This was the first breadbox built in Davis in recent years and provided the impetus for renewed research on breadbox heaters.

Performance

The Thigpen/Hunt house has worked very well for both heating and cooling. The backup gas heater has been left off for the last year and the Franklin stove has met all of the heating demand. The solar system and internal heat gain probably meet 75–85% of the heating requirement. Full cooling is also provided. The casablanca fan is used on the few days when the temperature exceeds 78°F. The breadbox water heater provides all of the hot water for nine months and significant preheating during the winter. The August 1978 utility bill for two included $1.63 for gas, $7.17 for electricity; the January bill was $4.18 for gas and $7.69 for electricity.

Improvements

Insulated shutters and drapes will further improve performance when they are installed. The house would actually work better with less glass. A water tank in the dining room would also improve the performance of the house and is being considered.

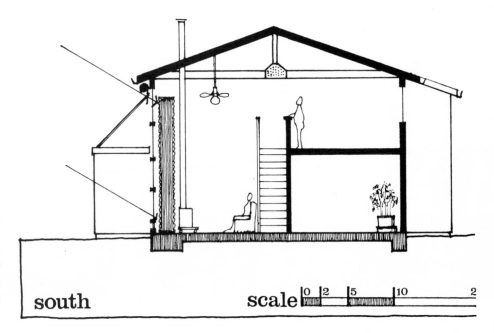

south

scale |0 |2 |5 |10 2

The Hofacre Residence

Design: *John Hofacre*
Construction: *John Hofacre*
Price Range: *(Adj. 1978)*
$41–45,000
Living Area: *1,200 sq. ft.*
Built: *1977*

John Hofacre thought a great deal about the design of his own house and it shows in the very efficient use of space and the delightful feeling of the house. It has two bedrooms as well as a section of living room that can be made into a third bedroom. The attic will be made into a studio. John put a wide, thin, window in the dining room for east viewing while seated at the table, and also for reducing the summer radiant heat gain. The horizontal window has been very popular and is now used on many other houses in the development. A covered deck on the northeast corner is very comfortable during the summer. An exterior utility room prevents summer internal heat buildup.

The System

This house is a simple water wall house. Heating is provided by 170 sq. ft. of south windows. Thermal storage consists of exposed tile in the living room, dining room, and entry, and 16 30-gallon drums filled with water. A Franklin stove and gas furnace are used for back-up heating. Water is heated in a thermosiphon system with a gas backup heater.

Performance

Despite single-pane glass, natural cooling has been adequate. Heating has been good and only about a half cord of wood was used for backup heating in the winter of 1977-78. Small portable electric heaters are used in the cold north bedrooms during the winter. The solar system provides full water heating in the summer and a boost in the winter. Utility bills for three in January were $12 and in August they were $8.50. Gas bills ranged between $1.50 and $4.

Improvements

One change that John will make is the addition of a security ventilation system. He might also add more mass, and use tanks instead of drums to make cleaning easier.

Narrow horizontal windows offer views to the east.

Painted, water-filled drums store the sun's heat.

Colorful supergraphic brightens north deck.

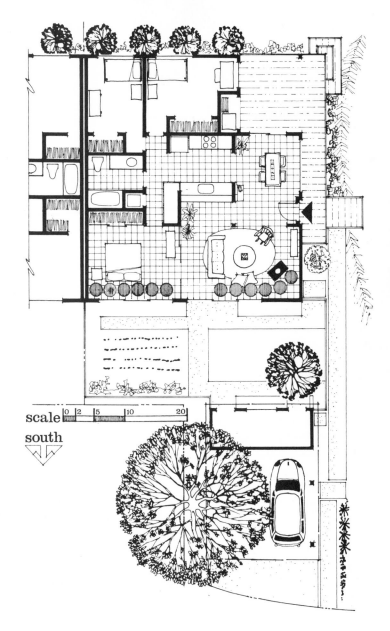

scale

south

85

The Piña Residence

Design: *Virginia Thigpen*
Construction: *Virginia Thigpen*
Price Range: *(Adj. 1978)*
$36–40,000
Living Area: *1,056 sq. ft.*
Built: *1977*

This commonwall house includes two bedrooms, one bath, living-dining room, and kitchen. A cathedral ceiling and an additional 100-square-foot loft make this relatively tight floor plan feel spacious and open. The Piñas are very pleased with the comfort and convenience of their house. A large deck to the south and patio to the north make excellent use of the outside yards. A planter box in the living room floor adds a nice touch of green to the inside.

The System

South windows are used for solar heating. The house has most of the slab exposed for heat storage and a water tank for added thermal mass. The storage tank is 3′ × 2′ × 8′ and holds about 500 gallons of water. The tank sits directly in front of the south windows and has wood veneer on the room side and a butcherblock top. Venetian blinds are used for privacy and help improve the windows' thermal performance. Backup heating is provided by a Franklin stove, and cooling is provided by cross-ventilation and a casablanca fan. A breadbox water heater will be added later.

Performance

The system provides full cooling for the house. The Piña house achieves 75–85% heating, with backup supplied by the wood stove.

Improvements

The house would perform better if insulated shutters and drapes were installed. However, it does so well now that it has been hard to justify adding them. The house should have security vents and fans added for automatic cooling when the owner is away.

Casablanca fan helps to distribute heat generated from wood stove and sun.

Counter water tank is covered with wood veneer and given a top surface to coordinate with rest of kitchen.

86

French doors open to a south-side deck of this commonwall house.

Water tank against window provides thermal mass and doubles as counter.

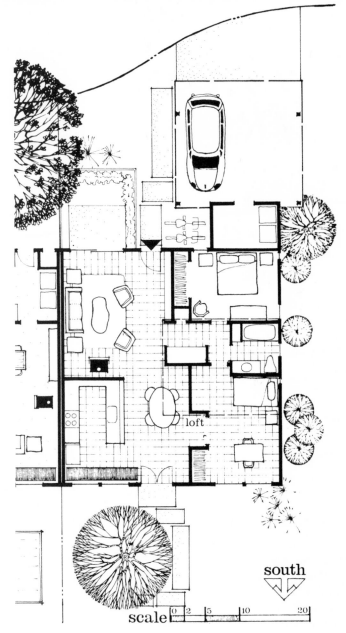

loft

south

scale 0 2 5 10 20

The Bainbridge Residence

Design: *Virginia Thigpen assisted by David Bainbridge*
Construction: *Virginia Thigpen*
Price Range: *(Adj. 1978) $30–34,000*
Living Area: *945 sq. ft.*
Built: *1977*

South side of these commonwall houses has generous amount of glass.

This commonwall house includes two bedrooms, bath, living-dining room, and kitchen. The cathedral ceiling and a 300-square-foot loft make it feel spacious and open.

David Bainbridge, developer of the tank wall concept, is particularly happy to live in the first house using this type of water wall. As he says, "Nothing helps people understand and believe in solar better than visiting my house on a cold winter day, when it's toasty inside, or a blazing hot summer day, when it's delightfully cool."

The System

The slab is painted dark brown and fully exposed for thermal storage. The water tank holding 1,000 gallons adds needed thermal mass. The tank is textured and painted to match the walls. Bookcases with 5-gallon water cans add extra mass to the bedroom. A Morsø stove provides all the backup heating. A casablanca fan is used for added cooling and a breadbox solar water heater on the carport provides all the hot water nine months of the year.

Performance

The system provides full cooling; temperatures typically stay below 80°F. The Bainbridge house will get 80–90% solar heating with utility bills averaging less than $5 per month for one person.

Improvements

This house will perform even better when the insulated shutters and drapes are installed on the windows. A security ventilation system for automatic cooling will also be added.

Water-filled cans support bookshelves, adding mass.

Water-filled tank absorbs sun's heat.

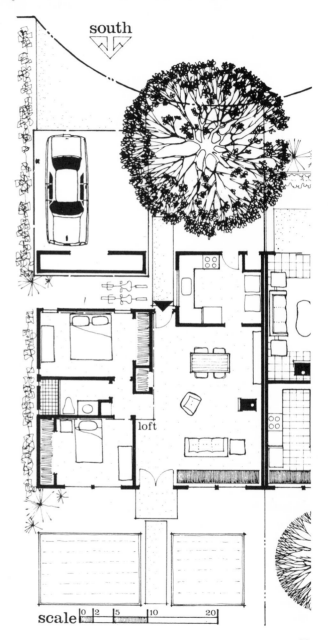

south

loft

scale |0 |2 |5 |10 |20

The Buckman Residence

Design: *John Hofacre*
Construction: *Mike Corbett*
Price Range: *(Adj. 1978)*
$63–68,000
Living Area: *1,800 sq. ft.*
Built: *1978*

The Buckman house has four bedrooms and two baths. Double rows of windows accommodate water mass under the lower set, while a view is afforded from the upper set. A deck on the top of the carport adds extra private space for basking in the sun or entertaining.

The System

The Buckman house has a water drum wall in front of the south windows on both floors. The water wall is in the living room on the bottom floor and the bedrooms on the top floor. These drums are painted black to absorb the winter sun and set behind a louvered wood counter. Tile floors on the slab in entry, dining room, and kitchen provide additional storage. This makes a much more attractive system than simple drums set in the space, and reduces the performance only a little. The windows open for cross-ventilation and cooling. Light colors, overhangs, and window placement help reduce the cooling load. A thermosiphon system is used for hot water heating.

Performance

The Buckman house gets full cooling and about 75–85% of the heating from its passive system. Comfort is reduced a little because the radiant temperatures of the drums are not felt by people in the room. The thermosiphon system provides all the hot water during the seven warmer months and much of it in the winter months.

Improvements

Insulated shutters or drapes would help the house perform better. A fully grown grape vine on the arbor will help shade the south windows and improve cooling.

South elevation with trellis for summer shade.

second floor

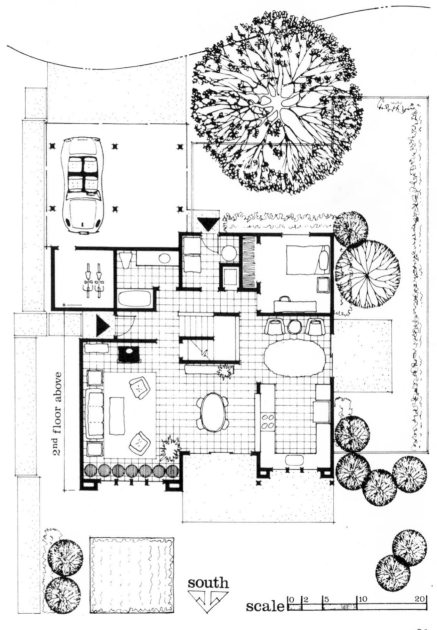

Louvered wall hides water-filled drums but still allows heat to radiate from them.

2nd floor above

south

scale 0 2 5 10 20

Clerestory

Basics

The clerestory house can offer many of the advantages that the simple passive house and water wall house can, along with better heat distribution. On the clerestory house, the roof is broken at the peak and a row of clerestory windows are added below the raised north roof. These windows provide good solar penetration to the north wall of the house in winter, yet can be fully shaded by the roof overhang in summer.

Clerestory houses are being used more frequently in Village Homes because they can provide very good performance with simple operation and very good reliability. The clerestory windows pose fewer weatherproofing problems than skylights do. In addition, part of the thermal storage can be moved to the north wall to provide balanced heating and cooling throughout the house.

The clerestory windows can also provide excellent ventilation if they are operable. The thermal stack caused by hot air rising from the ground level up to the roof peak at the clerestory provides a driving force for ventilation even when the sea breeze isn't blowing. The cathedral ceiling also lets warm air rise to the peak in the winter, and a small fan may be desirable to bring this warmth back to the ground level.

The openness and spaciousness of a cathedral ceiling point has proved very popular with the homeowners. The clerestory windows also provide beautifully balanced natural light to the house, which can also help reduce energy use.

Performance and Cost

The cost effectiveness of the clerestory house can exceed that of a well-done simple passive or skylight house if the windows and mass storage are adequately sized and properly placed. Most of the clerestory homes in Village Homes are undermassed for their glazed area and don't have insulated shutters or drapes for good glazing control. Nevertheless, they work well, are quite popular, and normally provide full cooling with occasional overheating and 60-70% of needed heating. With more mass and better glazing control, full cooling with no overheating and 90%+ heating would be possible.

The cost for the clerestory house is only slightly higher than that for the water wall houses. The added cost comes primarily from the increased complexity in framing, the added roof area required for the clerestory overhang, and the added windows. There will also be increased costs for the full cathedral ceiling, but these are used on many skylight, water wall, and simple passive homes, so this added cost is often not of concern. The addition of insulated shutters or drapes to the clerestory windows is also desirable. Both the windows and insulated shutters should be operable from the ground floor. The total cost should not exceed $500-1,000 more than for a water wall house and may be well worth it.

Operation

The operation of the clerestory houses is very similar to the water wall house and simple passive house. The only change is the addition of clerestory windows, which are operated much like the south windows on a simple passive house.

Maintenance

Maintenance is straightforward and should pose few problems. The only area where maintenance is more difficult than for a simple passive house is in cleaning and control systems for the clerestory windows and shutters. They are high and hard to reach to make repairs so they should be overengineered and well tested before installation.

Transferability

The clerestory house is almost universal in application and appeal. It can be equally at home in the hot desert or the cold north by simply changing the storage mass from water to concrete or rock to prevent freezing. Thanks to its self-ventilation from thermal stack effect, it is also good in the more humid climates where air movement is essential for comfort.

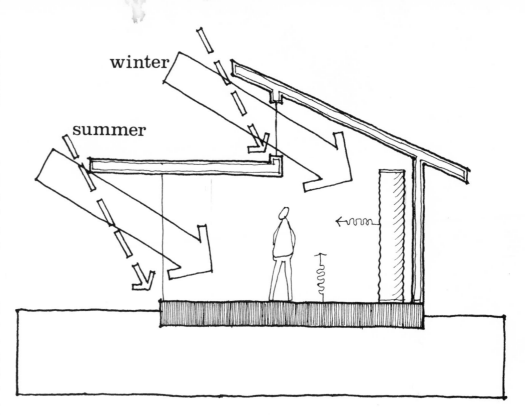

The Huestis/ Starr Residence

Design: *John Hofacre*
Construction: *Mike Corbett*
Price Range: *(Adj. 1978)*
$48–52,000
Living Area: *1,450 sq. ft.*
Built: *1976*

South features front flat roof and back-pitched roof for clerestories.

This house was designed for a student at the University of California at Davis who was involved in solar research. Gary Starr helped work out some of the details. Several solar research projects were undertaken while he and his solar-intrigued roommates lived here.[22] After he graduated, he left to start a solar consulting business in northern California and sold the house to Ed and Susan Huestis who moved in during the summer of 1978. The home has three bedrooms and two bathrooms with a den or playroom off the bedrooms. All rooms, except the baths, have south glass.

The System

The Huestis/Starr house has a flat, white gravel roof on the south side to reflect some of the winter sun into the clerestory windows. In addition, there are 220-square-foot windows on the south side of the house. Exposed tile is used for the floor throughout the house for storage. The bedrooms were located on the south so that the Starr solar group could also use their water beds for storage. But they soon tired of taking the bedcovers off every morning and installed water-filled culverts and drums to provide added mass. These were removed before the house was sold and have not been replaced.

The natural cooling features work very well. This house used "white" roof tile for additional reflectance, but this is

a bit too bright for neighboring residents. There is no west window and only five sq. ft. of east window.

Water heating is provided by a two-panel pumped collector built for test purposes by Gary Starr. A gas heater provides backup heating in the winter.

Performance

The Huestis/Starr house was extensively monitored and performed well. According to their study, full cooling and about 70–80% of the heating were provided by the passive system. Total utility bills were $13 in January for four adults, and $12 in July after the Huestis had moved in.

Improvements

The house worked well with the added mass of the water-filled culverts

Clerestories (at extreme right) add light to the kitchen area.

Clerestory runs the length of the house.

and drums, although even more would be desirable. Now that the mass has been removed there will be a decrease in the percent solar heating in winter, and in summer the house will be warmer in the afternoon.

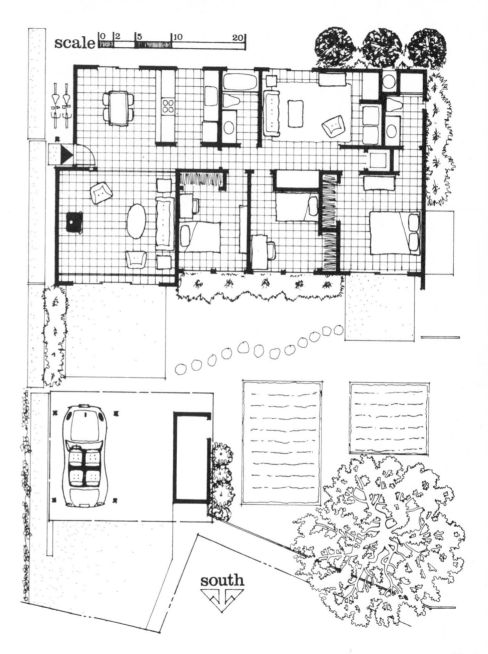

scale

south

The Bryant Residence

Design: *Mike Corbett*
Construction: *Mike Corbett*
Price Range: *(Adj. 1978)*
$60–64,000
Living Area: *1,560 sq. ft.*
Built: *1977*

East side windows capture creek view.

The Bryant house is a three-bedroom, two-bath house, which fits its site very well. A creek flows next to the house and visitors cross a quaint redwood bridge to get to the front door. The home has a very pleasing view to the south and west, and the window placement takes full advantage of it. Clerestory windows add natural light and a very pleasant warmth to the house.

The System

The 2 × 6 walls allow R-19 insulation to be used, and R-30 insulation is used in the ceiling. Double-paned glass has been used throughout the house. The sun enters through the clerestory windows and south windows to shine on tile floors in the entry, dining room, kitchen, living room, and on a three-foot tile strip along the south sides of the bedrooms. Each bedroom has an eight-foot-wide sliding glass door. Backup heating is provided by a wood stove and gas furnace.

Natural cooling is dependent upon the night sea breeze. Windows are shaded during the summer with over-hangs; light exterior colors are used and windows on the east and west sides are minimized. The Bryants have installed a wood and wire arbor in front of the west-facing windows to support grape vines. They have also used folding lou-vered doors on the windows to provide privacy while still catching the night breeze. A three-panel (72 sq. ft.) ther-mosiphon solar water heater provides the hot water in summer and preheats the 120-gallon water tank in the winter for a gas backup heater.

Performance

The Bryant house provides reasonably good cooling without any backup. In the winter, solar heating provides about 60–75% of the heating. The ther-mosiphon system provides full water heating in the summer and much of the heating in the winter.

Improvements

Additional mass and thermal drapes or shutters would be desirable. Temporary bamboo screens have been used on the west windows until the ar-bor can become lush with vines.

Path to the west-side front door.

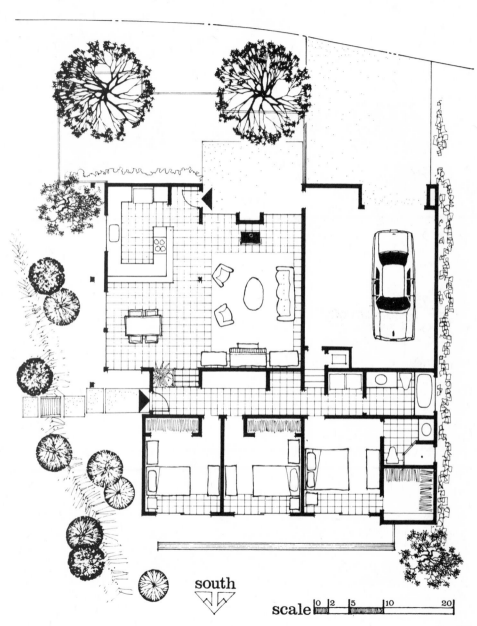

south

scale 0 2 5 10 20

Sliding glass doors admit sunlight to the south rooms; clerestory windows heat the north side.

Clerestory windows provide natural lighting in kitchen.

A wood-burning stove provides backup heat.

Wire arbor with young grape vines will shade south windows in summer.

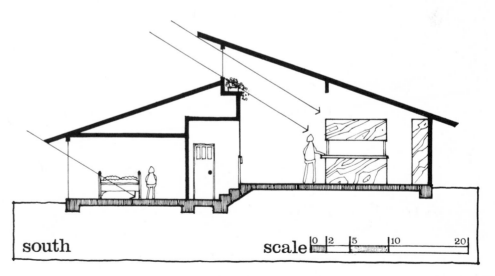

south   **scale** 0 2 5 10 20

The Dunn Residence

Design: *Mike Corbett*
Construction: *Mike Corbett*
Price Range: *(Adj. 1978)*
$44–48,000
Living Area: *1,278 sq. ft.*
Built: *1978*

Pat Dunn and her teenage children, Patty and Mike, enjoy the open feeling that the cathedral ceiling and clerestory windows add to the three-bedroom, two-bath house. Pat is developing a courtyard on the east side of the house off the dining room.

The System

This house is a more recent clerestory house, with tile floors in the entry, dining room, and kitchen for thermal storage. South windows augment the heating from the 50 sq. ft. of clerestory window. Only one room — a bedroom — doesn't have direct south exposure. Natural cooling is by natural cross-ventilation and induced ventilation through the windows and the operable skylight. South overhangs and light exterior colors also help keep the house cool. A Frontier airtight stove and a gas furnace are used for backup heating. Hot water is provided by a thermosiphon solar system with gas backup.

Performance

The Dunn house gets full cooling naturally with the induced cross-ventilation. Shading from trees and curtains is expected to alleviate the problem of local overheating from the east window in the future. The solar heating is expected to meet about 60–75% of the total heating demand. Total utility bill for a family of three for August was $14 and gas was $3.

Improvements

A rolldown canvas shade would probably be the easiest solution for the east-facing window. The house would perform better if it had more mass inside, preferably water. Insulated shutters on the clerestory windows would help in both winter and summer.

Cutout in stairs creates more visual space.

Overhangs shade south-facing glass in summer, but not in winter when sun is lower.

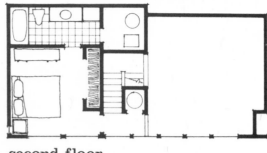

second floor

Sun through south glass heats tile floors.

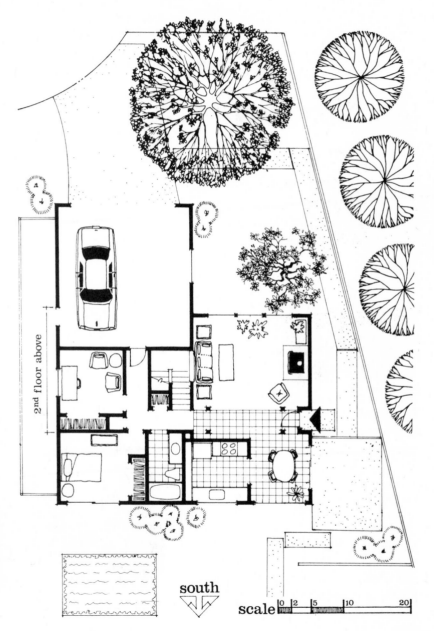

2nd floor above

south

scale 0 2 5 10 20

The Torguson Residence

Design: *John Hofacre*
Construction: *Mike Corbett*
Price Range: *(Adj. 1978)*
$56–60,000
Living Area: *1,554 sq. ft.*
Built: *1978*

Sue and Bruce Torguson and their two daughters live in this three-bedroom, two-bath home with a south courtyard. They particularly enjoy the second-story bedroom with its view and sun through the clerestory windows. The bedroom, master bath, and a deck are located upstairs to afford the parents ample private space of their own.

The System

The Torguson house has south exposure in every room. There is 64 sq. ft. of south glass in the upstairs master bedroom, 110 sq. ft. in the living, dining, and kitchen areas, and 55 sq. ft. in each bedroom. Storage is provided by the slab, which is tiled in the entry, dining room, and kitchen. Cross-ventilation from the cool sea breeze supplies cooling. Limited east and west glass, overhangs, and light exterior colors help reduce the cooling load. Backup air conditioning and forced air furnace are used. Hot water is heated with a 72-square-foot panel thermosiphon system.

Performance

The Torgusons use their backup air conditioning and heating to keep the house well within the 66–74°F range. Thus, they get less than full cooling and only about 60–75% heating. The thermosiphon provides much of the hot water for the year. Utility bills have been half the amount they used to pay in their former home.

Improvements

The Torguson house would benefit from added mass and insulated drapes and shutters. If mass and insulated drapes were added, the utility bills could probably be cut in half again, even with the same thermostat settings.

North-facing glass door sacrifices energy efficiency for aesthetics.

A cool upstairs sleeping porch is visible from the north.

second floor

South glass faces courtyard.

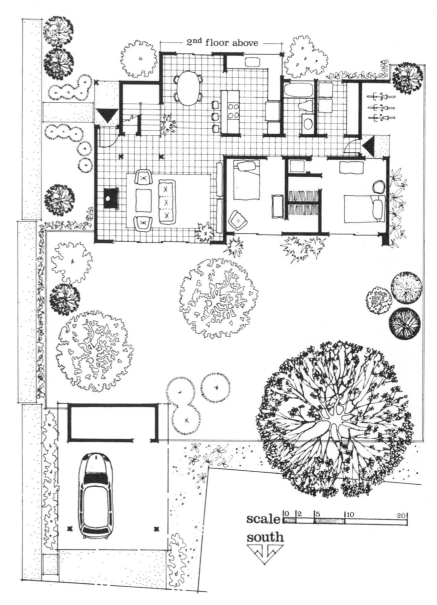

2nd floor above

scale | 0 | 2 | 5 | 10 | 20 |

south

Solarium or Solar Greenhouse

Basics

The solarium or solar greenhouse can be added to almost any house to provide solar heating. It consists of a room or space with south-facing glass, internal mass, and glazing control to provide shading in the summer and insulation in the winter. It offers low-cost space, much greater glazing area, and thermal storage, without making basic changes in the house shell. It also provides protection from overheating because it can be vented and isolated from the interior space.

The solar greenhouse or solarium can be a very attractive addition to a house and can provide much more than solar heating. It offers a protected and heated space to grow vegetables during the winter, helping to reduce energy use for energy-intensive purchased food, and enables plants to be started inside and transplanted outside for early maturation. Or it can be used as an added room or anteroom for dining or socializing.

At the present time there is only one solar greenhouse in Village Homes, but it is expected that more will be seen in new units of the development or as retrofits on existing dwellings if food prices continue to rise. They are less attractive in Davis than they would be elsewhere because the mild winters make it possible to grow many vegetables outside without additional protection.

Performance and Cost

The performance and cost of solar greenhouses are very difficult to generalize because they can vary more widely than any other type of passive solar house. In areas with hot summers they can be a liability unless they are either shaded, well vented and isolated from the living space, or shaded and well insulated with movable shutters or drapes. It is perhaps most fair to say that a solar greenhouse can provide a significant part of the heating demand and some of the cooling if the glass and mass are sized properly and the necessary shading, venting, insulated shutters and drapes are included. Cost may range from as low as $3 per square foot up to $35 per square foot. A good source of further information is *The Solar Greenhouse Book,* edited by James C. McCullagh (Emmaus, Pa.: Rodale Press, 1977).

Operation

The operation of a solarium or solar greenhouse depends in large part on its design and intended use. For a solar greenhouse that is designed for heating and cooling in Davis, the operation would be much the same as that described for a water wall house. In the winter insulated shutters or drapes are opened during the day and closed at night. Vents, windows, or doors are opened when the greenhouse is warmer than the house. During the summer, the shutters or drapes would be drawn during the day and opened at night. Vents and windows would be used to cool off the mass in the greenhouse at night. Or alternatively, the greenhouse can simply be shaded and vented continuously.

Maintenance

The maintenance of a solar greenhouse also depends in large part on its use. If it is used primarily for heating and cooling, it will require only installation of summer shades or growing suitable cover for an arbor and maintenance of insulating shutters and drapes. If, on the other hand, it is used for growing vegetables, fruits, or flowers the maintenance becomes much more involved and is far beyond the scope of this section. *The Solar Greenhouse Book* includes a full and comprehensive section on the maintenance of solar greenhouses.

Transferability

The solar greenhouse or solarium is particularly well suited for the colder areas of the United States where heating is a primary concern and where short growing seasons limit plant selection and reduce harvests. They are also easy to retrofit on most houses with good southern exposure; several very successful retrofit programs have been conducted across the country.[23] One of the very attractive features of solar greenhouses for retrofit is the simplicity and very low cost of construction possible using recycled materials.

Owner-builder adds glazing to his attached greenhouse.

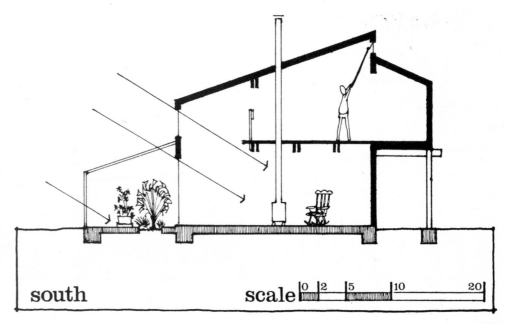

south   scale

The Morgan Residence

Design: *Bob Morgan*
Construction: *Bob Morgan and friends*
Price Range: *(Adj. 1978) $28–32,000*
Living Area: *1,000 sq. ft.*
Built: *1978*

Bob Morgan both designed and built this delightful little house. As partner in a local fix-it shop and a former architecture student, he used the many skills he has developed to the fullest. Without having to pay for labor, he kept the total cost, including the lot, below $25,000. The simple floor plan with few walls and use of the natural concrete slab as a finished floor also helped keep the cost down. It is a very spacious-feeling house with a kitchen and eating nook, living room, and a southern greenhouse on the bottom floor; and a bedroom, loft, and bathroom on the second floor. A deck is incorporated off the second-story bedroom, and Bob uses it nightly during the summer as a cool, breezy place to sleep or look at the stars.

Diagonal cedar siding makes an attractive exterior material, and the house is topped with a light-colored, concrete tile roof. Summertime shade is provided by deep blue morning glories over the south arbors. These are in delightful contrast with the orange marigolds in the south garden.

The System

This house is a simple passive house with a solar greenhouse. It uses the greenhouse and south windows to collect the winter sun. The greenhouse has an earthen floor so it can be used directly for growing vegetables or sprouting beds. A door leading to the nook area can be opened and heat will rise, traveling through the house. Storage is provided by the slab now, but water containers will be added when time and money allow. Night cross-ventilation cooling is aided by a two-story arbor covered with morning glories and use of very few windows on the east and west. During the summer the greenhouse is vented to prevent collection of heat. A breadbox solar heater will be in-

Deciduous morning glories grow on a trellis, providing summer shade.

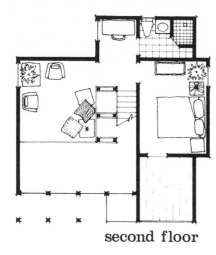

second floor

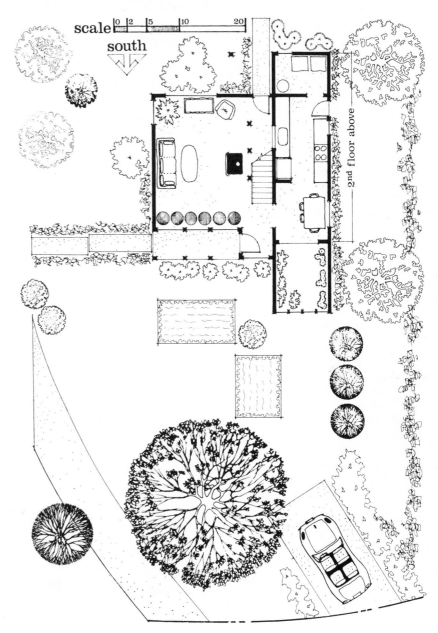

south

2nd floor above

stalled for hot water. The only backup space heating is provided by an airtight Morsø wood stove.

Performance

The house provides full cooling with only a few warm afternoons and nights. The solar system currently provides probably 70-80% of the heating and the stove provides the rest. Hot water will be heated by gas until the solar system is installed.

Improvements

The major improvements already planned include adding water storage,

Sun through south glass warms concrete floor.

108

A shade trellis defines path to entry.

insulated shutters and drapes, and a breadbox water heater. An insulated drape or exterior shade for the south wall of the house would also be helpful.

The greenhouse will heat up substantially in the winter since it has little thermal mass. Opening the greenhouse directly to the house for a long period of time will tend to overheat the house. Also, the house has a relatively high amount of exterior surface for its small size, and this will hurt its thermal performance.

Windows are used sparingly on the north.

Skylight

Basics

The skylight house has much in common with the simple passive house and the water wall house, but adds a new area of solar glazing — skylights in the roof. Typically the skylight houses include at least two 4 × 6 skylights. These skylights allow sun to penetrate more deeply into the house and enable water containers to be placed further back in the house. This helps overcome the difficulty the simple passive and water wall houses have in heating rooms on the north side.

The skylight houses have been built often in Village Homes because they are fairly easy to build and attractive. The skylights also allow good solar gain with privacy. The skylights encourage stargazing from the living room on clear spring and fall nights and provide a nice warm house interior on cold winter days. They also allow some of the water storage to be relocated to the center of the house and free more of the south windows for viewing.

However, there are also some difficulties involved in adding large skylights. First, the skylight must have a very well insulated shutter with a tight seal. A fairly good shutter has been developed, in large part by Dave Norton and Mike Corbett, but it remains cumbersome and reasonably expensive. Shutters also allow some summer heat gain, and exterior shade panels are now being used on some of the skylight houses for better solar control. And finally, the addition of a skylight, particularly to a tile roof, makes flashing for waterproofing more difficult.

Performance and Cost

The performance of the skylight house should be about the same as the clerestory house if the glazing and mass are appropriately sized and located. Most of the skylight houses in Village Homes are slightly undermassed for their glazing, and therefore don't achieve optimum performance. However, they work well and provide full cooling and 75-85% of the heating.

The cost of the skylight houses is a little higher than that for the clerestory house. This increase is due to the skylights and their insulating shutters. The total cost of these on a typical skylight house is $1,500-2,000. This added glazing, however, allows better placement of mass and more gain in winter and can be cost effective if used carefully. The shutters do require diligent operation by residents during winter months and often are not used as effectively as they might be.

Operation

The operation of the skylight house is very similar to that described for the simple passive house and the water wall house. In the summer the windows are opened at night to allow the sea breeze to cool off the interior mass. If an exterior shade is not placed on the skylight, then the skylight shutter can be opened to allow some radiant and convective cooling to occur through the skylight. In the winter the skylights, window shutters, and insulating drapes are opened during the day to allow the winter sun to strike the storage mass, tubes, tanks, or concrete or tile floor, and they are closed at night.

During the spring and later in the fall when temperatures are fairly mild during the day and at night, the skylight shutters can be left open all the time. This allows the occupants to enjoy the view of the stars and moon, as well as the puffy clouds and clear skies of spring and fall.

Maintenance

The maintenance of the skylight house is also very similar to the simple passive house and the water wall house. Insulating shutters should be checked for wear and adjusted periodically. Exterior shade panels, if used, must be installed in the spring and removed in the fall. They might need new paint every few years.

Weatherproofing around the skylights may need to be checked in the fall, and additional sealing or flashing repairs should be made if necessary. The skylights can be cleaned at the same time to allow maximum solar gain in the winter. Although a layer of dust

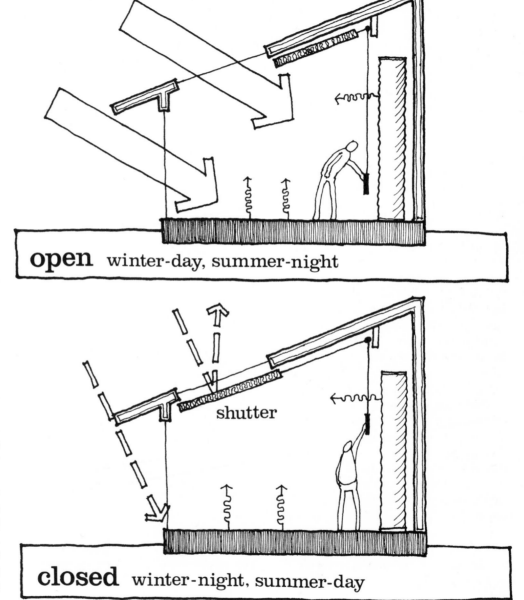

open winter-day, summer-night

shutter

closed winter-night, summer-day

will reduce transmission only 10-15%, it can be a very important 10-15%. Cleaning is well worthwhile if fall rains don't naturally wash the windows.

Transferability

The skylight houses are readily transferable to most other climate areas in the United States. They are perhaps best for areas without much snowfall, unless a fairly steep roof pitch or gambrel roof is used to keep the skylights free of snow. Masonry should be used for storage in place of water to reduce freezing problems in colder areas where water tanks might freeze.

The skylight houses are particular-ly well suited for areas with a high percentage of diffuse rather than direct solar radiation. In Seattle, for example, where much more of the radiation is diffuse, the skylight house would perform better than a simple south wall house. It would also, as you might expect, be very important to weatherproof the skylights extremely well.

In hotter areas the skylights can be of added value if they are operable, to increase natural ventilation. A summer exterior shading device will prove desirable in many of these areas to reduce unwanted heat gain. If this is designed properly, the skylight could be kept open even during a rain storm.

Counterweight eases pulley operation.

Wood shutters roll on garage door tracks.

Pulleys and wire for shutters can be run through walls.

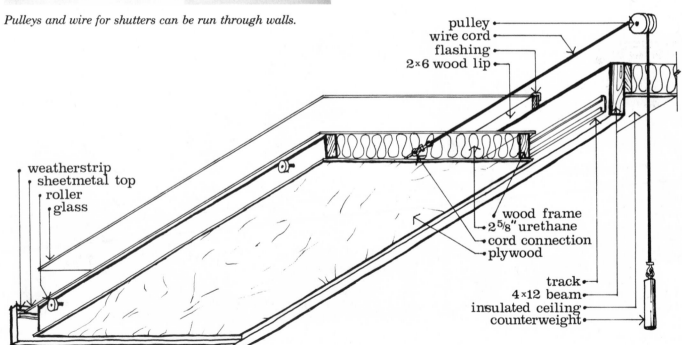

pulley
wire cord
flashing
2×6 wood lip

weatherstrip
sheetmetal top
roller
glass

wood frame
$2\tfrac{5}{8}''$ urethane
cord connection
plywood

track
4×12 beam
insulated ceiling
counterweight

The Thayer Residence

Design: *John Hofacre and Mike Corbett*
Construction: *Mike Corbett*
Price Range: *(Adj. 1978) $56–60,000*
Living Area: *1,573 sq. ft.*
Built: *1977*

The Thayers and their young son enjoy their four-bedroom, two-bath, two-story home. The south yard is fenced and includes a patio, while the backyard is open and allows a nice view of the Thayers' vegetable garden.

The System

The Thayer house has three skylights to augment the windows on the south side of the house. The skylights have movable shutters with R-19 insulation. Storage is provided by the tile floor on slab in the dining room and kitchen. Three water-filled culverts are also included in the center of the house where the winter sun from the skylights will strike them. The backup heating is done by a wood stove or forced-air gas furnace.

Natural cooling is provided by overhangs on the south side for full summer shade, window placement for cross-ventilation, and use of light-colored exterior walls. A fan is used upstairs for improved ventilation and cooling.

The water heater is a thermosiphon solar system with an 80-gallon storage tank. It uses a gas backup heater in the winter.

Performance

The Thayer house stays comfortable in the summer without air conditioning, and only limited use of the backup heater is required in the winter. The backup water heater is only needed in the winter.

Improvements

The house has worked well, but the addition of exterior shades for the skylights, and on west and east windows would be helpful. Insulated drapes or shutters for the windows would also help the house perform better. A lighter roof may slightly improve cooling in the summer.

Natural plantings and ground covers eliminate need for high-maintenance lawn.

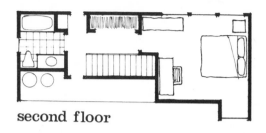

second floor

Skylights in open position provide heat and light.

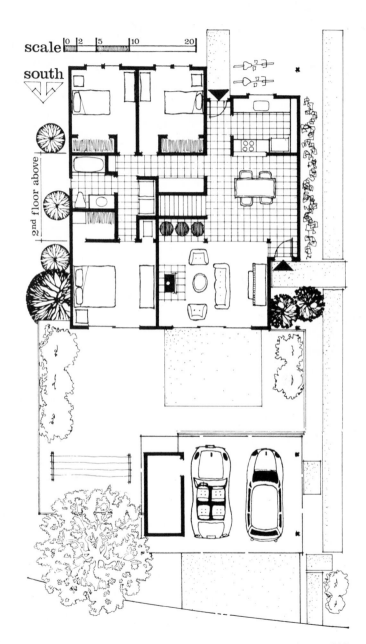

scale

south

2nd floor above

The Pearlman Residence

Design: *Mike Corbett*
Construction: *Mike Corbett*
Price Range: *(Adj. 1978)*
$62–64,000
Living Area: *1,560 sq. ft.*
Built: *1976*

The Pearlmans' three-bedroom, two-bath house has a private master suite on the second story with a deck off the bedroom.

The System

Their house uses skylights to collect the sun and the tile floor and three 7′ × 21″ tubes for storage. Cooling is by natural ventilation. Water is supplied by a thermosiphon solar system with an 80-gallon gas backup.

Performance

The Pearlman house provides 75–85% solar heating in the winter and does well in the summer with only natural cooling. The windows on the east side cause some local overheating. The January utility bill for two people was $30 and the August bill was $18.

Street-side fence provides privacy on the south side of the house.

Water-filled culverts and tile provide thermal mass.

second floor

Open skylights allow sunlight in.

Improvements

Insulated shutters, exterior summer shades on the skylights and east glass, less north and east glass, and a wood-burning stove would improve performance.

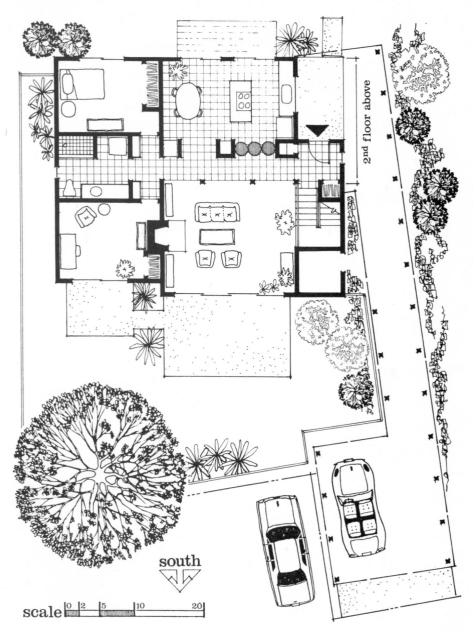

2nd floor above

south

scale 0 2 5 10 20

The Thigpen Residence

Design: *Virginia Thigpen and John Hofacre*
Construction: *Virginia Thigpen and Mike Corbett*
Price Range: *(Adj. 1978) $45–49,000*
Living Area: *1,147 sq. ft.*
Built: *1975*

This house has a very interesting and efficient plan with two bedrooms, bath, and a neat little loft. The thermal performance is increased by being commonwall with another house. It has a built-in planter box with a tree, *Ficus benjamina,* that is now reaching towards the cathedral ceiling. Windows are placed to take full advantage of this lovely greenbelt corner lot. Wood parquet floors add a feeling of warmth to the floor. Walls are constructed of 2 × 6's with R-19 insulation, and the roof is also very well insulated. Planter boxes in the backyard provide vegetables and greens most of the year. A small courtyard in front includes a clothesline.

The System

This house was one of the first skylight houses in Village Homes and has one 4′ × 6′ skylight near the top of the ceiling. It lets sunlight reach far into the house and onto the loft. Storage is provided in the slab (the parquet reduces performance only a little), 30-gallon drums full of water, and a double layer of brick on a divider wall. Backup heating is provided by a wood stove and gas heater. The cooling cross-flow-ventilation is aided by a vent, with a closable door near the top of the vaulted ceiling. An exterior shade of bamboo and an interior insulated panel are placed on the skylight in the summer. Solar water is provided by a two-panel, horizontal tank, thermosiphon system with an 80-gallon storage tank in the attic. Gas backup is provided for use as necessary.

Performance

The solar heating contribution is about 75-85% with much of the rest provided by wood. The solar water heater provides full heating in the summer and much of the heating in the winter. An instant hot water tap can be used to provide hot water for tea or coffee and to augment warm solar water for dishwashing. Total utility bills were $16 in January, $7 in August, and gas bills ranged between $1.50 and $11.

Improvements

The addition of thermal drapes or shutters to windows and some type of security ventilation would be desirable. Virginia would also use rectangular water tanks instead of drums if she were to build it again. The windows are not very tightly sealed, and better windows would be helpful. An arbor to the south would provide desirable shade for the south windows and a place for additional grapes to grow.

Skylight encourages lush plant growth.

118

South roof of this commonwall house supports skylight (on left) and four collectors (on right).

Water-filled drums support a desk top.

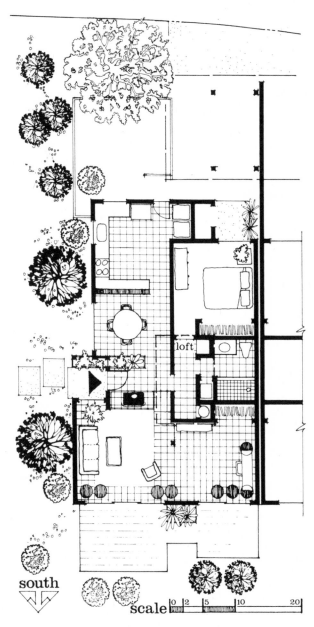

loft

south

scale   0   2   5   10   20

The Gallant Residence

Design: *Mike Corbett*
Construction: *Mike Corbett*
Price Range: *(Adj. 1978)*
$79–83,000
Living Area: *2,080 sq. ft.*
Built: *1977*

Jim Gallant's house is very spacious with four bedrooms, two baths, and a laundry and crafts area off the kitchen. Hanging cabinets visually screen the kitchen from the living room and dining room. A large, round, stone fireplace is central to the open area. Two metal culverts are tucked away into wing walls on either side of the living room and screened with louvered blinds. A pleasant stone patio and trellis complement the street-side courtyard.

The System

Mass is provided by the tile floor, two 10' × 24" culverts full of water, and a massive stone fireplace. Heating back-up is by central fireplace and furnace.

The house makes very effective use of cross-ventilation and is shaded for cooling. A thermosiphon system pro-vides the hot water with the storage in a chimneylike structure.

Performance

The skylight system used in this house provides full cooling and 70–80% of the heating. The thermosiphon system supplies all of the hot water in the summer and much of it in the winter. Total utility bills were $28 in January and $17 in August. Gas bills ranged from $4 to $16.

Improvements

Insulated drapes would be desirable. An airtight stove would be better than the fireplace. North-facing glass could be reduced.

Owner examines solar water tank located on roof of the house to take advantage of thermosiphon principle.

Sliding glass doors look out to private, south courtyard.

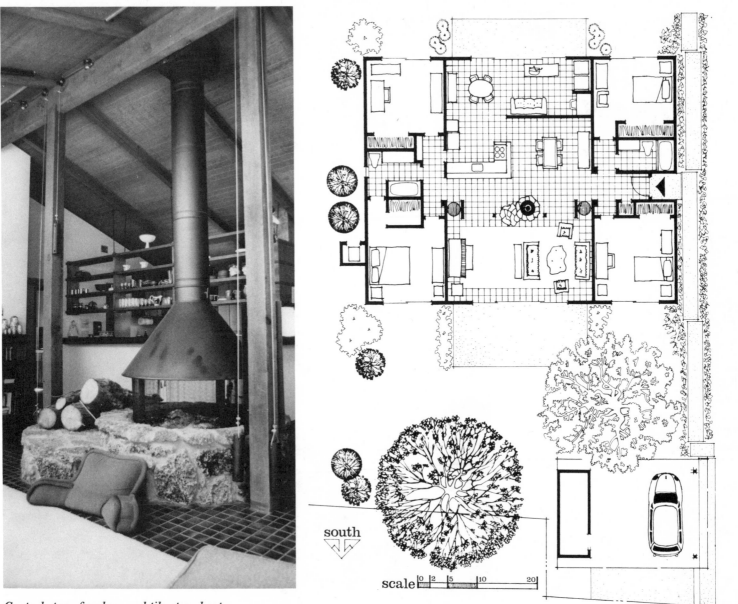

Central stone fireplace and tile store heat.

south

scale 0 2 5 10 20

The Andrews Residence

Design: *Mike Corbett*
Construction: *Mike Corbett*
Price Range: *(Adj. 1978) $61–65,000*
Living Area: *1,700 sq. ft.*
Built: *1977*

The Andrews house is a three-bedroom, two-bath, skylight home. Cabinetry is incorporated in the spacious entry hall, creating an interesting first view for the visitor. A fireplace with raised, tiled hearth is located between the living, dining, and kitchen areas, creating a dramatic focal point.

The System

The Andrews house has three skylights high on the roof and south windows to collect the winter sun. Because the skylights are higher on the roof than most other houses, the sun can penetrate more deeply into the house. Thermal storage is provided by exposed tile floor throughout the house and by six 12' × 21" water-filled culverts. Backup heating is provided by the central fireplace and a gas furnace. Sliding shutters cover the skylights, but the rest of the windows have only conventional drapes. Cooling is by cross-ventilation at night with careful shading and control of summer sun by window placement and light exterior colors. An operable two-foot-square skylight placed above the culverts at the ceiling peak releases summer heat. A pumped hot water heater, with a gas backup, provides hot water.

Performance

Full cooling (with a few warm afternoons) is provided by the system, and 75–85% of the heating demands are met. The thermosiphon system provides most of the warm season water heating as well as much of the winter heating.

Improvements

Insulated drapes or shutters would help improve comfort in the summer and heating percentage in the winter. Exterior shades over the skylights in the summer would also help improve cooling. The north bedrooms are sometimes cold, and the house should have a duct and a fan to carry warm air from the living room into the bedrooms. An airtight wood-burning stove would be more efficient than the fireplace.

Raised, tiled hearth adds to house's simple lines.

Culverts and tile provide thermal mass.

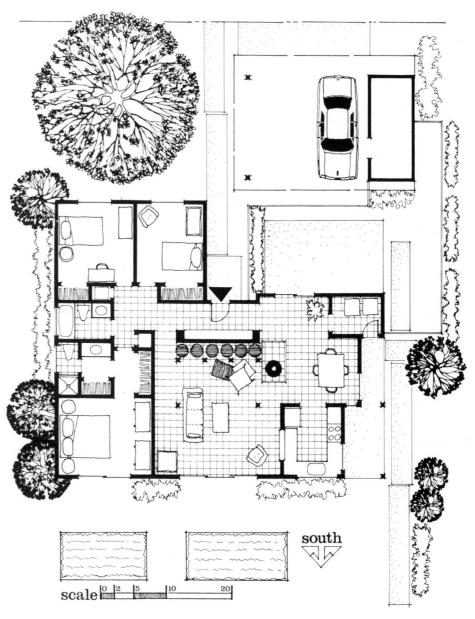

scale 0 2 5 10 20

south

The Corbett Residence

Design: *Mike Corbett*
Construction: *Mike Corbett*
Price Range: *(Adj. 1978)*
$93–97,000
Living Area: *2,100 sq. ft.*
Built: *1977*

South view into courtyard reveals skylights and collectors.

The Corbetts, coplanners and developers of Village Homes, live in this 2,100-square-foot home with their two children. The home has three bedrooms, two bathrooms, a small office, loft, and drafting room alcove. It provides quiet spaces for the children and parents, and a large living room which is often used for homeowner meetings. The loft is an especially cozy place for the family to gather on winter evenings. A north-side deck is an appealing location for summertime meals, and a deck off the upstairs bedroom is a cool place to sleep on a summer night. A special hinged door allows the bed to roll out onto the sleeping deck. The large lot is used intensively with a large garden, chickens, orchard, and courtyard.

The System

The Corbetts have chosen a passive solar skylight design, developed by Mike, which covers most of their living room ceiling with glass. The room has the appearance of a light, airy greenhouse. Five insulated panels slide down to cover the skylights on winter nights and summer days. Tile is used on most of the bottom floor to add mass. Six water columns line a wall on the bottom and upper stories, providing 1,000 gallons of water for additional heat storage.

The Corbetts' only backup heat is provided by two wood-burning stoves. One stove, located in a hallway, has metal ducts that passively carry the heat to the north side of the house.

Water heating is provided by a thermosiphon system and stored in a 120-gallon tank. A 20-gallon electric water heater may be manually switched on to heat water during rainy or foggy periods. In addition, water coils inside one of the stoves can be used to heat water whenever the stove fire is burning. Good security ventilation, light-colored stucco, and a concrete tile roof, as well as a southern exterior grapevine overhang help keep the house cool on summer days.

Performance

The house keeps within the 65°-70°F range most of the winter, and in summer natural cooling has kept the

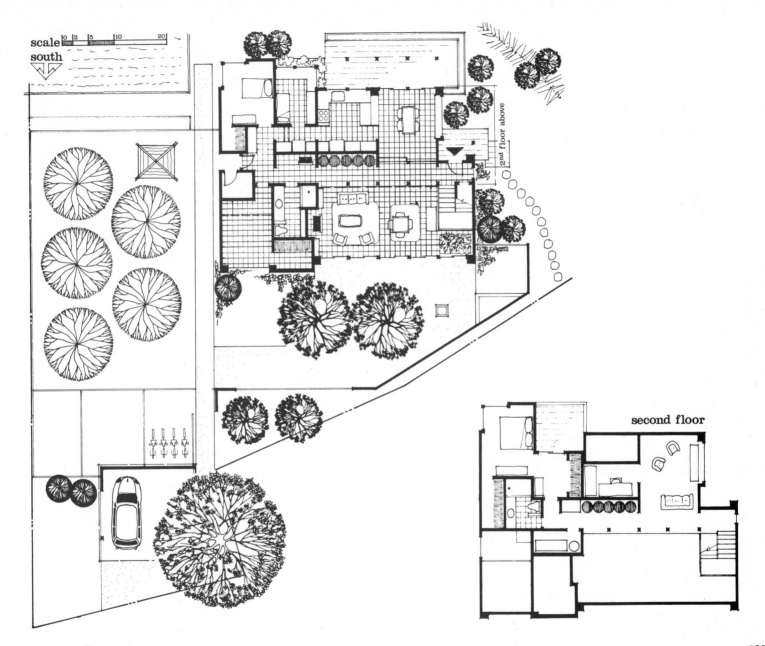

scale

0 2 5 10 20

south

2nd floor above

second floor

house below 83°F. The Corbetts are both convinced they would not be happy returning to a regular house—one which is not "naturally" comfortable.

The Corbetts find they use their backup water heater an average of five months of the year for about an hour a day.

The house is all electric. Monthly utility bills have ranged between a summer low of $9 and a winter high of $22.

Improvements

The Corbetts found that in the summer it was necessary to cover their east-facing windows and cover their skylights with Masonite, painted white. These measures may prove less important once the surrounding vegetation grows to its full height, and subsequent heat gain is reduced. A casablanca fan would also aid the comfort level in the summer.

Shutter on tracks above staircase is positioned over skylights.

Heat through south-side glass is absorbed by tile floors and culverts.

An airy feeling is created when the skylights are open.

Ladder leads to child's loft.

Wood arbor accents garden walkway.

The Okusako Residence

Design: *John Hofacre*
Construction: *Mike Corbett*
Price Range: *(Adj. 1978)*
$50–54,000
Living Area: *1,374 sq. ft.*
Built: *1975*

This three-bedroom and two-bath house utilizes space quite effectively. The Okusakos like to come home in the evening to enjoy the attractive view to the southwest with its ever-changing light. Their backyard has a special feature not usually found with a subdivision house — a turtle pond. And the turtles they raise would probably also comment favorably on the house and the site design.

The System

The Okusako house has four skylights with insulated shutters to augment the solar gain of the south windows. The thermal mass storage includes an exposed tile floor and four 7′ × 21″ culverts holding 2½ tons of water. Good cross-ventilation and shading, including a patio covering with movable lattice, help provide cooling in the summer. Hot water is supplied by a two-panel thermosiphon flat plate solar heater.

Performance

The systems provide full cooling, although the house overheats occasionally in the summer. The solar heating system provides about 75–85% of the required space heating. A Frontier wood-burning stove and gas furnace with set-back thermostats are used for backup heating. The solar water heater provides all the hot water for 8–9 months and supplemental heating the rest of the year.

Lattice on south patio can be removed to maximize penetration of winter sun.

Two solar collectors and four skylights on the south-side roof.

Skylights open with easy pull on weighted cord.

Improvements

This house would perform better if the windows all had insulated shutters or drapes. Summer cooling would be better if exterior shades were installed on the skylights and west-facing windows. Additional mass would also be desirable.

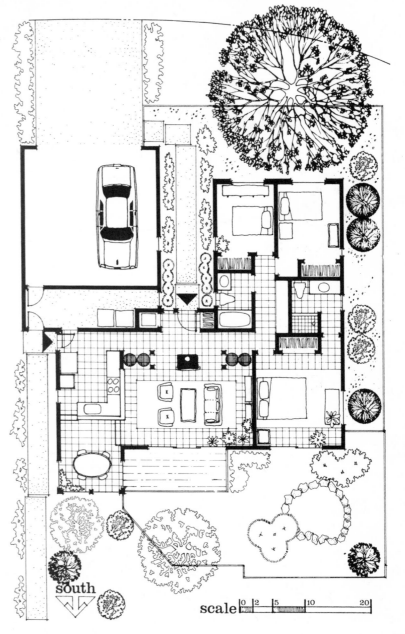

south

scale 0 2 5 10 20

The Olsen Residence

Design: *Mike Corbett*
Construction: *Mike Corbett*
Price Range: *(Adj. 1978)*
$39–43,000
Living Area: *1,098 sq. ft.*
Built: *1978*

Tonya Olsen, a graphics designer who drew many of the plans for this book, lives in this home with her two teenaged sons. The house has three bedrooms and one bath. Tonya has turned a corner of the kitchen into a design studio. The living room opens onto a large wood deck enclosed by the street-side courtyard. It provides an appealing place for summer sunbathing. The large skylights provide a delightful spot for basking in the winter sun.

The System

The Olsen house is one of the more recent skylight houses. The tile-covered slab and six water-filled culverts (10′ × 18″) provide storage for sun entering the skylights and sliding glass doors on the south side of the house. A 100 cfm fan operated by a switch draws warm air from the living room ceiling and carries it through a duct into the north bedrooms. This provides good heat distribution through the house and solves the problem of cold north rooms. A wood stove and gas furnace are used for backup heating. Cross-ventilation through north and south windows provides cooling. Cooling loads are reduced by south overhangs, light exterior colors, and few windows on the east and west sides. A thermosiphon solar heater with gas backup is used for hot water heating.

Performance

The Olsen house provides full natural cooling with only a very few warm afternoons and evenings. The passive features should meet about 75–85% of the heating demand with the wood stove meeting much of the rest. The thermosiphon system will provide all of the hot water in the seven warmer months and half of the hot water in the winter.

Improvements

Insulated shutters and drapes on the windows would be helpful for both heating and cooling. Exterior shade panels for the skylights would help reduce cooling demand. Vegetation will help keep this new home cooler next summer. Natural lighting in the kitchen, via a ceiling skylight, would reduce the lighting demand.

Unshuttered skylights add light and warmth.

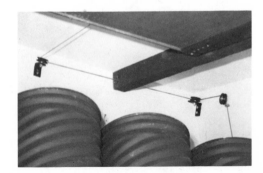

Pulley detail for operating skylights.

South glass with deck faces courtyard.

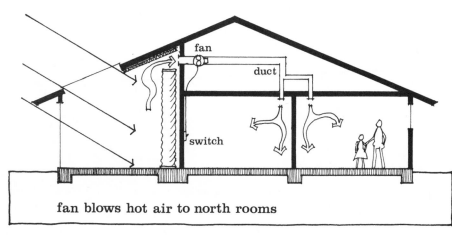

fan blows hot air to north rooms

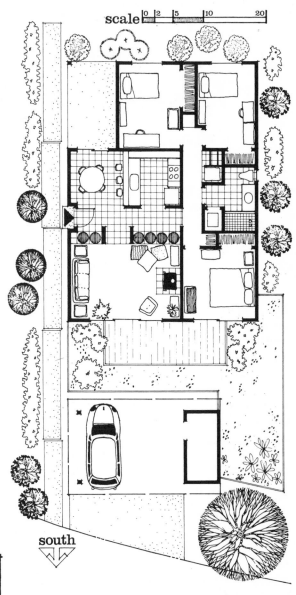

scale 0 2 5 10 20

south

The Woods Residence

Design: *John Hofacre*
Construction: *Mike Corbett*
Price Range: *(Adj. 1978)*
$30–34,000
Living Area: *814 sq. ft.*
Built: *1978*

The Woods bought one side of this neat little two-bedroom, one-bath commonwall house. They liked the open-space floor plan in the house and the low price. The south yard with patio is fenced and very private while the backyard opens onto the lush greenbelt. Culverts form a wall separating the living room from the kitchen.

The System

The house relies on south windows and two skylights to collect the winter sun. This heat is stored in the tile floor, which is used everywhere except in the bedroom. In addition, four 10′ × 21″ water-filled culverts are used for added storage. A wood stove is used for backup heating. Natural cooling is provided by good cross-ventilation, shading, and light-colored walls and roof. Trellises have been added in the south yard for extra shade in the summer.

Hot water is heated by an 80-gallon pumped system with electric backup. A pumped system was used because the collectors were mounted high to match the roof on the Maeda house next door. A thermosiphon would not work with this collector location.

Performance

The house has worked reasonably well although it has been a bit warm on some of the hottest summer days. It hasn't been through a winter yet, but with the system it has it should work quite well, providing 75–85% of the heating.

Improvements

This house would be improved if exterior shades were made for summer shading of the skylights. Insulated shutters or drapes on all the windows would be desirable. A gas backup hot water heater would be much more economical than the electrical backup, but it would be more difficult to turn off and on manually.

Open skylight in living room.

Culverts act as room divider.

Note flat plate hot water collector on roof.

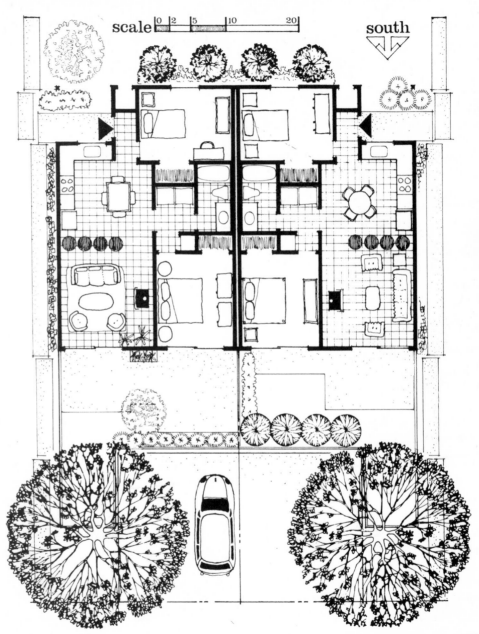

scale 0 2 5 10 20

south

133

Suncatcher

Basics

The suncatcher house is similar to the clerestory house but has a reflective south roof sloping down to the clerestory rather than away from it. This reflective roof and a reflective soffit or eave on the north roof act as a funnel to increase solar gain to the house. In summer the roof angles prevent direct sun from entering the house although it continues to collect diffuse sun, and the window must therefore be backed with an effective insulated shutter.

The suncatchers focus this sun on the top of a series of water columns in the center of the house. Heat and coolth are transferred to other parts of the house by radiation and convection, and fairly good thermal balance is achieved if an open plan is used. If an open plan is not used and south rooms are separated from the mass wall, there must be sufficient mass to prevent overheating from their south glass.

The suncatcher design was specially developed by Jon Hammond of Living Systems for use on lots where south wall exposure was not possible. In Village Homes it has been used only on lots with good south wall exposure at this time, but it is one of the few solutions for lots where south exposure is a problem. As built, the suncatcher requires complex framing and poses weatherproofing problems that both increase the cost and may increase maintenance problems.

The suncatcher houses have interesting interiors and work well. The interiors are open and visually interesting, and the exterior shape proclaims "solar" in loud terms. For those with extra money to spend and a willingness to

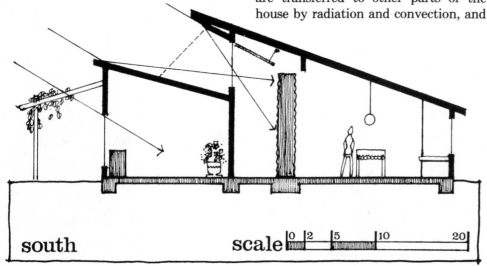

south   scale 0 2 5 10 20

provide careful maintenance for the roof, they provide an intriguing alternative to the clerestory house.

Performance and Cost

The suncatcher houses have performed well. The balance of mass to glazing has been better than most of the other houses, and full cooling and 75-85% heating have been realized. Minor overheating problems in south rooms were rectified by adding additional thermal mass.

However, the suncatchers have proved difficult to weatherproof. These problems result from the funnel shape, which captures rain as effectively as it captures sun. Despite careful detail work by the builders, leaks are difficult to stop. In part, these may be caused by a high pressure pocket created by wind on the funnel shape that is strong enough to drive water even upward through cracks in the glazing or roof. It should be possible to eliminate or minimize these problems using different roof materials or flashing details.

The increased complexity of the house adds substantially to the price. Nonstandard roofing materials are used, and as a result, additional costs are incurred. The weatherproofing is a major concern and requires more work and different materials than a standard roof. With experience gained on the first models, the costs on more recent houses have gone up, rather than down, and would probably run $1,000-2,000 over a similar size clerestory house with equal mass.

Operation

The operation of a suncatcher is virtually identical to that required for a clerestory house. Shutters and drapes are closed on winter nights and summer days. Windows are opened on summer nights to allow the sea breeze to cool the interior mass.

Maintenance

Work may be required to maintain waterproofing. In addition, the windows and insulated shutters can be as complex as those on a clerestory house and will require periodic maintenance. Insulated drapes and shutters should be carefully tested before installation.

Transferability

The suncatcher would best be used on lots with poor south exposure in mild climates. In other areas, the clerestory, water wall, or simple passive house will cost less and work almost or equally as well.

Reflective aluminum roofing bounces additional sunlight into clerestory.

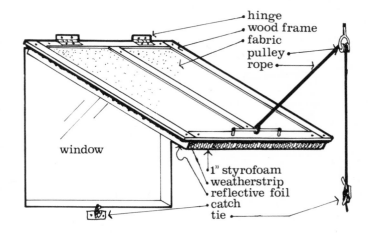

hinge
wood frame
fabric
pulley
rope

window

1" styrofoam
weatherstrip
reflective foil
catch
tie

The Maeda/Nittler Residence

Design: *Living Systems (Jon Hammond, Jim Plumb); Bruce Maeda*
Construction: *Virginia Thigpen*
Price Range: *(Adj. 1978) $65–69,000*
Living Area: *1,600 sq. ft.*
Built: *1977*

The Maeda/Nittler house was designed for maximum solar heating and cooling to satisfy Bruce Maeda, a passive solar expert, and his wife Lynne, a teacher. It has three bedrooms, two baths, an inviting loft, and a delightfully open living/dining room with a bay window that includes bench seats.

The System

Natural heating and cooling are supplied by the south windows and suncatcher roof. The row of eight 10′ × 20″ water columns are used for thermal storage. Reflective, insulated shutters are provided in the clerestory windows.

Cooling is provided by good cross-ventilation through the house and good shading from an arbor to the south. Hot water preheating is provided by a breadbox solar water heater with gas backup.

Performance

The house is being extensively monitored, and in the winter of 1977 the solar system by itself met 60% of the heating demand with the thermostat set at 65°F. If the temperature were allowed to drop to 58°–60°F at night and the internal gains are considered part of the passive system (they really are), then the house would probably meet about 75–85% of the heating demand. Total utility bills for a family of three were $9 in August and $17 in January. Gas bills ranged from $1.50 to $9.

Improvements

This house would work better if it had insulated curtains or shutters on the windows. The house has suffered leakage at the base of the suncatcher. Caulking and sealing have controlled the leaks, but a different window flashing detail would solve the problem permanently.

Shutters are aluminized to reflect clerestory light onto culverts.

Lower south roof is also reflective to maximize solar radiation through clerestories above it.

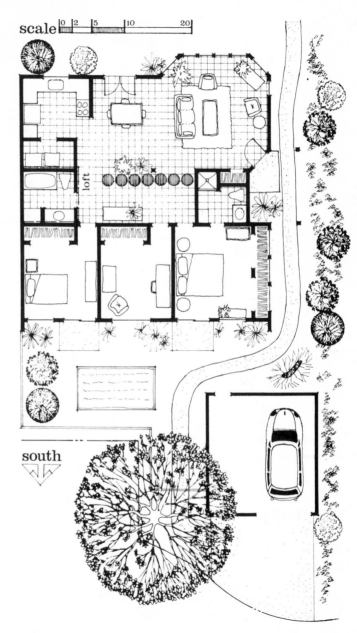

scale 0 2 5 10 20

south

Culverts divide hall from living room.

The Drach Residence

Design: *Living Systems (Jon Hammond, Greg Acker)*
Construction: *Jay Davison, Virginia Thigpen*
Price Range: *(Adj. 1978) $65–69,000*
Living Area: *1,600 sq. ft.*
Built: *1977*

The roof angles with reflective surfaces focus light into windows.

Steve Drach worked closely with his architect, and the resulting one-bedroom, one-bath plan is very interesting and especially fitted to his needs. A very open living room, dining room, and study with wood stove occupy the south and west sides of the house. The bedroom and bath are to the north and isolated for privacy. The living room wall is decorated with a large mural of the night sky.

The System

The Drach house uses aluminum roofing on the lower roof and the soffit to collect the winter sun and direct it through the windows onto seven 10-foot-high water-filled columns. Insulated shutters are installed on the clerestory windows so that they may be closed off on winter nights and summer days. The slab is used for added thermal storage. Cooling is provided by cross-ventilation at night, overhangs, and use of light exterior colors. A breadbox solar system is used for hot water heating.

Performance

The Drach house provides full cooling and about 75–85% heating. The breadbox water heater provides all of the hot water in the summer.

Improvements

Insulated shutters and drapes on all the windows would be desirable and could improve even more the solar heating and natural cooling. A wider overhang on the south would help shield the south wall from winter storms.

A striking graphic emphasizes the lines of this living room.

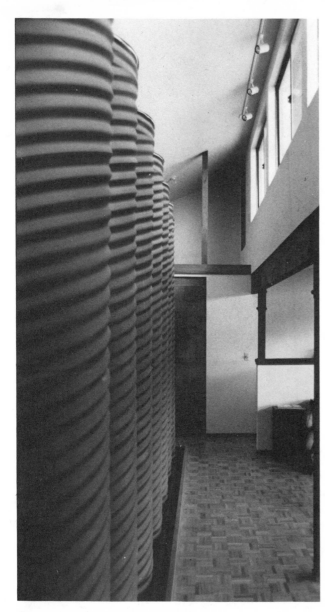

Sun strikes water-filled culverts.

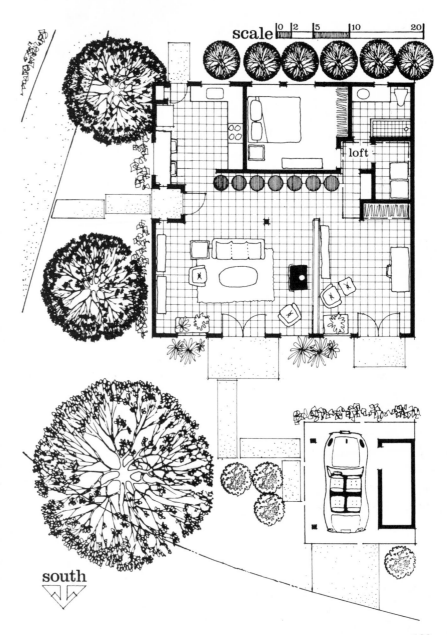

scale

loft

south

Passive/Active

(Using floor slab for storage)

Basics

Several passive houses in Village Homes use an active solar system to supplement the passive heating. This system was conceived by Mike Corbett. A flat plate collector on the roof provides the additional collector area for heating. A differential thermostat controls a pump that pumps an ethylene-glycol water mix through the collectors to polybutylene pipes in the slab where the heat is stored. This extra heat is used to supplement the heat collected and stored in the house by its passive solar design.

This type of system has been used to make simple passive houses more fully "solar" without adding more passive storage or window area. It also provides more carry-over and better control. The extra heating is added to the north side of the house and this helps balance the heating. The slab must be thermally exposed for the system to function well. Such systems would be even better in areas that are colder than Davis, where heating alone is vital, but the houses do provide better comfort than a simple passive house at an added cost of about $2,000. This is comparable to the more expensive passive houses and would be a particularly good investment where cooling is less important.

Performance and Cost

The performance of these passive houses with supplemental active systems is very good for heating and they may provide 75-85% of the heating demand. The high storage to collector area ratio keeps operating temperatures down and enables the system to operate very efficiently. These systems also avoid storage losses common to most active systems. Cooling is similar to that realized with simple passive systems.

Operation

The operation of a passive house with active assist is similar to that for the simple passive house with one addition — turning on the system in the fall by using a switch. From there operation will be automatic and this is one of the attractions of this type of system. The ethylene-glycol water mixture is heated in the flat plate collector and pumped to the slab where it circulates through a series of polybutylene tubes to heat the slab. The system may work too well on a hot winter day and it may be necessary to turn it off or open the windows to cool off a bit. The system is turned off manually in the spring.

Maintenance

Maintenance is also very similar to that for the simple passive house, with minor service required for the active system, pump, and controls. Rain usu-

Valve controls are neatly hidden.

Hall rug is narrow so that tile can radiate heat from hot water pipes underneath.

ally washes the collectors in the fall, but if it does not, they should be washed for maximum performance. The controls should probably be tested every year or two to ensure that they are working properly and turning the pumps on and off at the proper time. The water should be checked for corrosiveness, and occasionally replaced.

Transferability

These passive houses with active assist will be attractive for use in many climate areas. They are particularly effective in areas that are colder and require limited cooling. This includes much of the United States, and they should be carefully considered in many of these areas. They offer particular value for vacation home use where a passive house would not do very well unless it were fully automated. The active system can keep the temperature up until users arrive to add the passive performance.

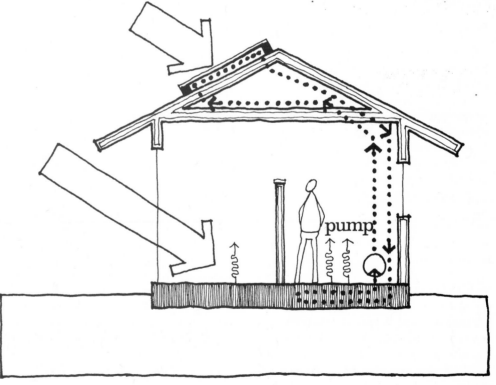

The Johnson-Musso Residence

Design: *Mike Corbett*
Construction: *Mike Corbett*
Price Range: *(Adj. 1978)*
$53–55,000
Living Area: *1,441 sq. ft.*
Built: *1978*

Southside displays south glass, trellis and six solar collectors on low roof.

Bud Johnson, his wife Sharon Musso, and their small son moved into their three-bedroom, one-bathroom home in June and are very pleased with it. The living room and the master bedroom open onto a fenced, street-side courtyard. The extensive south glass and high ceiling in the living room give the room an airy, spacious feeling.

The System

There are 180 sq. ft. of south windows in the living room and six 12′ × 21″ water-filled culvert tubes forming a back wall. Full tile floors in the entry, dining room and kitchen, and strips of exposed tile elsewhere are also used for

Tile floors and culverts provide thermal storage in living room.

thermal storage. The house has three flat plate solar collectors (78 sq. ft.). Backup heating is provided by a wood stove and gas furnace. Cooling is provided by good shading and by a large trellis on the south. Hot water is supplied by three additional flat plate collectors using a thermosiphon system with a 120-gallon storage tank and a gas backup.

Performance

The passive/active system provides full cooling and most of the heating. Between 70–80% of the heating should be supplied by the solar system. The solar hot water supply should meet all the domestic water needs for seven months and 50% for the three coldest months.

Improvements

The addition of an exterior shade for the second story south living room windows would help improve comfort levels in the summer. Insulated shutters or drapes would be helpful for both heating and cooling. The numerous hard, high mass surfaces provide a noise control problem which could be reduced with acoustical ceilings and textured fabrics.

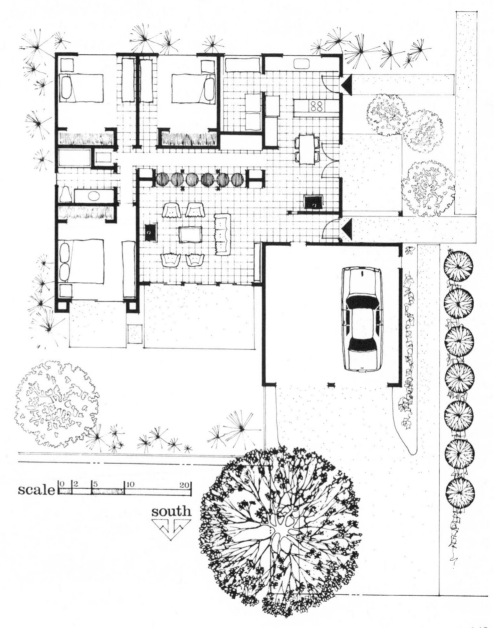

scale |0 |2 |5 |10 |20|

south

143

The Toft Residence

Design: *John Hofacre*
Construction: *Mike Corbett*
Price Range: *(Adj. 1978)*
$40–44,000
Living Area: *1,014 sq. ft.*
Built: *1978*

Kathy Toft and her roommate have been very pleased with this two-bedroom, one-bath house. The floor plan has worked out very well, and the minimal hall/passageways and open plan make the house seem large even though it is quite small. The dining and living areas are lined with sliding glass doors which open onto a soon-to-be landscaped yard and greenbelt. Carpeted corridors are lined with tile on either side to assist radiant heating from the slab. A long, narrow dining room window affords an additional view to the west while minimizing heat gain and heat loss.

The System

This is a house with a passive/active system. It includes two flat plate collectors (48 sq. ft.) which pump water directly to pipes set in the north side of the slab. South windows provide passive heating for the south side of the house with the tile-covered slab used for storage. A gas forced-air heater provides backup heating. Cooling is provided by good cross-ventilation at night, careful solar control with light exterior colors, few windows on the east and west, and south overhangs.

Performance

The house has stayed reasonably cool through the summer without a backup system. There isn't any experience with winter heating yet, but it should be in the 70–80% range with the added collectors and good heat transfer to the slab. The thermosiphon system should provide 70–80% of the hot water over the year.

Improvements

Insulated drapes and shutters would be desirable. Additional thermal mass in water containers in the south rooms would also be helpful. Security latches allow the sliding glass doors to be locked while still open a few inches. A fan would help pull the cool night air through these small openings at a faster rate, cooling the home more rapidly.

Roof has two panels for space heating and two panels for water heating.

The bedroom is edged with tile, facilitating radiant heating of this north room.

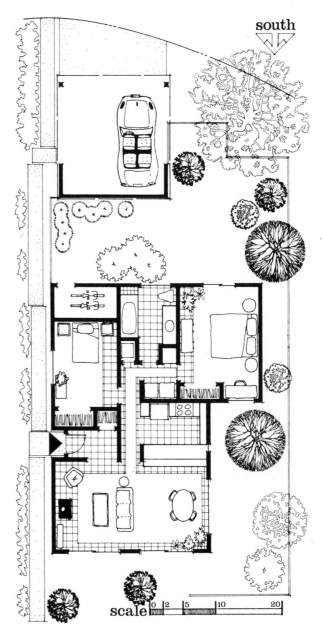

south

scale   0 2   5   10   20

145

Passive/Active

(Using a water wall for storage)

Basics

In addition to the houses that use an active system to assist the passive systems with storage in the floor slabs, there are houses where the active and passive systems are combined. This particular system was conceived by Marshall Hunt. It uses an active system coupled to a water wall to provide cooling and heating. The system includes flat plate collectors, a pump, and a differential thermostat to determine when to pump water from the collectors to the tank.

The active system uses fairly standard equipment and design. The water wall is a very large water tank built with sheet steel, much like the Bainbridge-Piña tanks. This tank stands much taller and has internal braces to keep it from bowing out. Because water is circulated through it, there is an epoxy coating on the inside. As it is plain steel, welding is easy and the required fittings and junctions are simple to install. Once the tank becomes more than two times as high as it is wide, some form of bracing is required, and the tanks used in Village Homes have been tied into the frame of the house.

The active system uses automatic draindown to prevent freezing. The hot water runs directly through the large tank. The cold water inlet line also has a heat exchanger loop in the tank. Supplemental cooling on hot summer days is provided by running all the cold water used in the house through this loop.

This type of house was developed in Village Homes and insofar as we know has never been used before. It is a clever combination and should prove extremely comfortable and efficient. It looks very promising for use in many other areas of the United States as well.

A forklift moves a large, metal water tank into the foundation of a new house.

Performance and Cost

This type of house should work very well in Davis. It will easily provide full cooling and 75-85% of the required heating. The large storage element is exposed to the space, providing the radiant surfaces lacking in the previous passive houses backed with an active system. The use of this large storage mass (1,500 gallons in one house) will also keep operating fluid temperatures in the flat plate collector low. This increases the efficiency of the collector by reducing losses in both the collector and the transfer system.

The large exposed mass will also make nighttime cooling by cross-ventilation very effective. In addition, valves are included to route the cold water used through a heat exchanger in the tank in summer which will augment the already very good ventilation cooling.

The cost is not as high as might be expected because the passive storage doubles as the active storage. Total cost of this type of system should be similar to that for the passive house backed with an active system, or from $3,000-5,000. Because this type of house augments cooling as well as heating, it may prove to be a better investment.

Operation

This type of house is operated like a simple passive house with two added features. Incoming cold water is routed throughout the heat exchanger in the tank in the summer and windows are opened at night in the summer for cross-ventilation. In the winter insulated drapes and shutters are closed at night and opened during the day, and the differential thermostat is turned on, allowing the pump to circulate when the water in the collectors is warmer than the water in the tank.

Maintenance

The pump, collector, and controls do not require maintenance. The pump and control systems should last 10 years and will require replacement when they fail. Insulated shutters and drapes may also require minor maintenance.

Transferability

This type of system is very attractive for many areas of the United States. The only drawback to very widespread use is the cost, but it should prove well worthwhile. The added heat from the roof collector can make it very effective even in cold areas, and in hot areas the cooling could be backed up using a supplemental passive cooling technique.

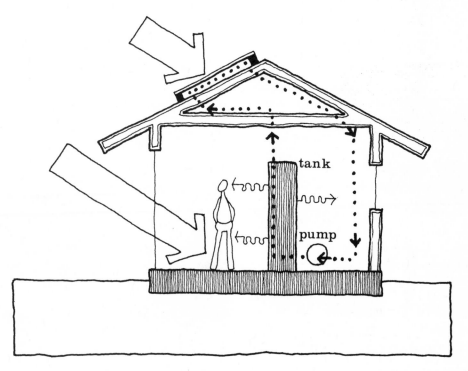

The Springer Residence

Design: *Jay Davison,*
Sydney and Fred Springer
Construction: *Jay Davison*
Price Range: *(Adj. 1978)*
$60–64,000
Living Area: *1,600 sq. ft.*
Built: *1978*

The Springers wanted a spacious house that would be solar heated. They got it with this design, inspired by Marshall Hunt and developed by Jay Davison. A cathedral ceiling combines with open living spaces on the first and second stories to create a delightful feeling of spaciousness. The stairway leads your eye up to the second story where a study-den provides a view of the lower rooms. A clerestory window adds natural light to the study, making it a very appealing work area. A welded steel tank, 2′ × 10′ × 12′, is so well integrated into the structure that it is hardly noticeable except for the tank top which extends into the study-den.

The System

The centrally located 1,800-gallon rectangular tank forms the heart of the heating/cooling system. What is unique about this system in comparison to the other water wall systems is the addition of two solar collectors sloped at about 60° that are mounted to one side of the clerestory windows. Water is circulated from the storage tank through the solar collectors using a small pump, controlled by a differential thermostat. It functions much the same as a solar water heater except that the water stays in the tank and radiates heat directly to the interior of the house. Because the tank is set back from the south side of the house so that it can exchange heat with most of the living area, it does not receive very much direct radiation. Therefore, the collectors make up the difference. Since the storage volume to collector area ratio is very large (about 37 to 1!), the collectors operate at a very low temperature, and therefore a very high efficiency. If the tank is heated

Collectors on upper left corner south roof.

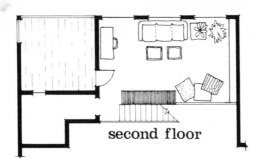

second floor

Because of its size, the water tank must be installed during the construction process.

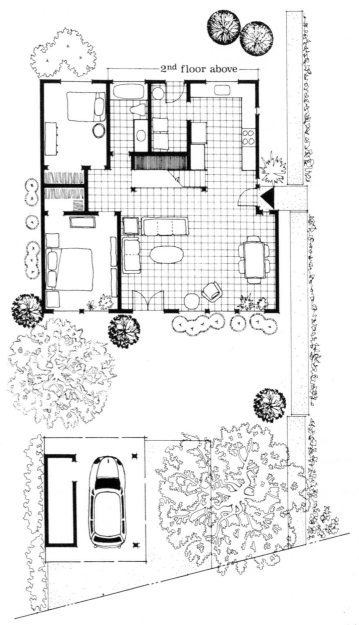

2nd floor above

When house is complete, the painted tank resembles a half wall.

A staircase along the tank.

only to 80°F, it will store and put out a substantial amount of heat.

The large tank also provides summer cooling by losing heat to the cool night air. Circulation is aided by the north and south windows. This cooling is supplemented, probably to a small extent by running all of the cold water used in the house through a 120-foot copper heat exchanger immersed in the tank. The heat exchanger can be bypassed during winter operation.

Just below the collectors, which are used for heating the interior of the house, are a pair of collectors that provide hot water for domestic use. The two sets of collectors give the visual impression of an open book. The water heating system is of the pumped type. Pumped recirculation of warm water from the storage tank is used to prevent freezing of the collectors during cold weather. If a power failure were to occur, a valve would open which would allow the tank to reverse thermosiphon to the collectors for complete protection against freezing. Freeze protection for the upper set of collectors is not a problem since they drain into the 1,000-gallon storage tank when the pump stops. Both the heating and domestic water heating systems were designed and installed by Natural Heating Systems.

Performance

This hybrid, or "passive/active" system, should provide a substantial portion of the heating and cooling for the house. Because of the large exposed surface area of the tank, it should exchange heat with the air very well and will also "see" a large part of the interior of the house and exchange heat radiantly with it. Full cooling should be provided and 75–85% heating. A wood stove and gas furnace are available to provide supplementary heating. The pumped-flow water heater should meet 65–75% of the yearly hot water demand.

Improvements

The only major improvement that suggests itself is the addition of insulated drapes and curtains. If the collector area provided for heating proves to be too small, space is available for the addition of more collectors.

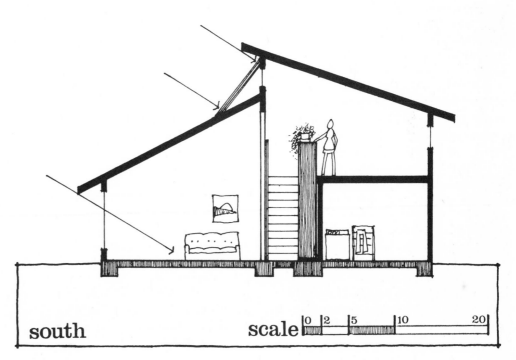

south   scale 0 2 5 10 20

151

Active

Basics

There are also several houses in Village Homes that are predominately "active" solar houses with passive features. They use flat plate collectors on the roof to collect the sun's energy and a pump to move it to an insulated storage tank and then on to the floor slabs as heat is needed in the house. These are called active systems because they collect energy in one place and transfer it to another area for use with auxiliary energy.

They are more expensive than most of the other solar houses, but they can provide better heat distribution than most of the passive houses. They also require less owner involvement and are more dramatically "solar" than passive houses. Their major drawback is cost, which is equal to or more than that of the most expensive passive systems. For this added cost heating performance is about the same as the expensive passive house with better distribution. Cooling remains similar to that of the simple passive house.

Performance and Cost

The active solar houses provide very effective heating with good heat distribution. The active houses with passive assist in Village Homes work well and provide 75-85% of the required heating. Their cooling capability is similar to the simple passive houses.

Costs for the collectors, pipes, pumps, controls, insulated storage tank, and delivery system run about $6,000-7,000 installed, including hot water. This does provide the comfort of a radiant heating system with less manipulation required than a passive system. If the cost of a conventional radiant system and solar water are subtracted from the system cost, then the actual cost is only $3,000-4,000, or about the same as the more expensive passive systems.

The active systems used at Village Homes are less costly and more efficient than the typical active system. These improvements are achieved through use of radiant heat distribution, the use of low cost, appropriately chosen materials, and the lack of complexity in the design and operation.

The radiant heat distribution used in conjunction with the active systems offers several advantages. Since the concrete slabs have considerable mass, their capacity for storing heat is great. This heat capacity can be harnessed with radiant heating. The result is a smaller and less costly storage tank than would be required with forced-air distribution. Secondly, solar collectors can collect more heat if they are operated at a lower temperature. Due to the large surface area of radiant heating panels and the nature of radiant heat, water at a temperature of less than 90°F can provide adequate heating. Thus, less collector area is required.

There are a few drawbacks related to radiant heating. Response time is generally very slow since the entire slab must be heated before heat is delivered to the living space. One system uses a timer which turns the system off early in the evening and on again early in the morning, thus providing an automatic night setback. Having the slab exposed for radiant heating is a plus for natural cooling since it allows the slab to act as thermal storage in summer as well as winter.

Copper or iron pipe has traditionally been used to conduct hot water in radiant heating systems. These materials are expensive and subject to corrosion, and a leaky pipe in a concrete slab can have disastrous consequences. After considerable research on the subject, David Springer, Mike Corbett, and another Davis developer persuaded the city to allow polybutylene plastic pipe to be used in radiant heating applications. Polybutylene is capable of carrying water at a temperature of 180°F and a pressure of 100 psi, is not subject to corrosion, is easier to install, and costs less than copper or iron. Though it has a lower thermal conductivity, closer spacings between the piping runs (9 inches instead of 12) provides equivalent heat output with more uniform temperature at the surface of the slab.

Operation

Operation of all three of the ac- tively heated houses is basically the same. When the collectors become hotter than water in the storage tank, a differential thermostat turns on a pump that fills the collectors with water from the storage tank. Water is circulated between collectors and storage until the collectors become cooler than the stored water, at which time the pump stops and the collectors are allowed to drain back into the storage tank.

Two room thermostats are used to control the distribution of heat to the floor, one to distribute solar heated water through the piping imbedded in the slab, and one to switch the system to backup heating if the water in the storage tank is not hot enough to maintain the indoor temperature at the desired level. A conventional water heater is used for backup heating. The dual thermostat feature allows the resident to increase system efficiency by lowering the setting on the backup thermostat.

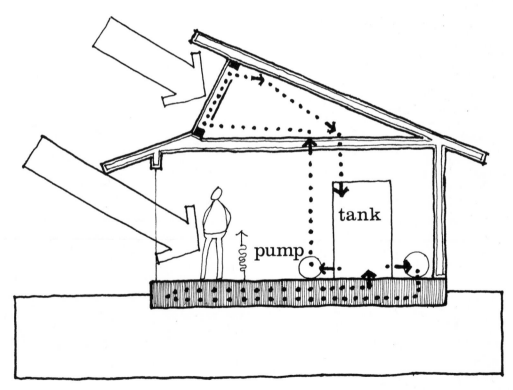

Domestic hot water is heated by passing it through a copper heat exchanger immersed in the storage tank before it is delivered to the conventional water heater. A bypass allows the solar heated water to be used directly during summer months so that the water heater can be turned off.

Transfer of heat to individual rooms is controlled by valves which can be opened to increase the radiant heat output, or closed to restrict it. Each room or zone is controlled by a separate valve.

In addition to bypassing the water heater, the only seasonal adjustment that is needed is the closing of the control valves. If the valves are left open during the summer, some hot water will flow to the radiant panels by convection. Otherwise, the systems are completely self-regulating.

Maintenance

Little or no maintenance is required. The two pumps used are water lubricated and should provide trouble-free service for many years. Since the storage tank is not tightly sealed, some water may be lost by evaporation, and addition of water to the tanks is necessary about once a year.

Transferability

The active systems of the type used in Village Homes are practical anywhere where construction on a concrete slab is possible. They are particularly ef-

fective in colder climates where the expansive glazing required for passive heating would produce substantial heat loss or require extensive shuttering, or

where solar exposure or architectural limitations prevent the use of passive heating.

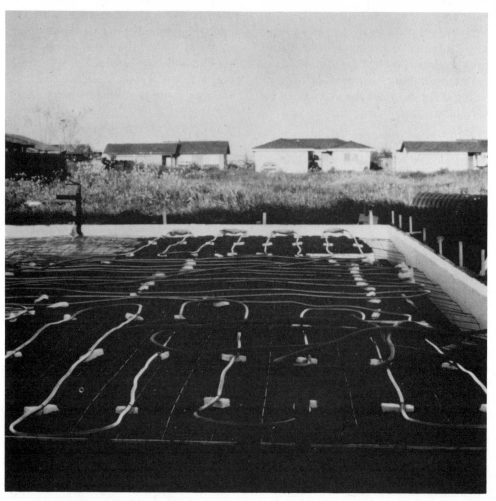

Copper piping is laid before the concrete floor is poured.

154

Solar water storage tanks are well insulated.

The MacGregor/ Garbini Residence

Design: *John Hofacre*
Construction: *Mike Corbett*
Price Range: *(Adj. 1978)*
$66–70,000
Living Area: *1,510 sq. ft.*
Built: *1977*

Ian MacGregor and Susan Garbini worked very closely with the designer in the development of this two-story, three-bedroom, two-bath house. The house has a large, fenced east yard with a spacious redwood deck.

The System

The MacGregor/Garbini house utilizes eight panels on the 60° angled roof for solar collection and a 260-gallon storage tank for thermal storage. The heated water is pumped through copper coils in the concrete slab foundation and a two-in. lightweight slab on the second floor. A copper heat exchanger passes through the tank to heat or preheat the domestic hot water backup. This is

helped with passive solar heating through south and clerestory windows with storage in the tiled concrete slab. Backup heating is provided by a 40-gallon gas water heater and wood stove.

Natural cooling by cross-ventilation is aided by overhangs on the south, light exterior colors, and minimal glass on the east and west.

Performance

The house works quite well for heating and cooling. The total gas bill for space and water heating was less than $30 from October through May 1978.

Improvements

Insulated drapes and shutters and

added thermal mass would help improve the performance. The solar collector's 60° angle causes glare problems in the homes of neighboring residents for short periods of time during spring and fall.

Pipes laid in thin concrete slab to provide radiant heat on second-story floor.

Courtyard with shaded east glass.

156

second floor

Second-floor fireplace and windows.

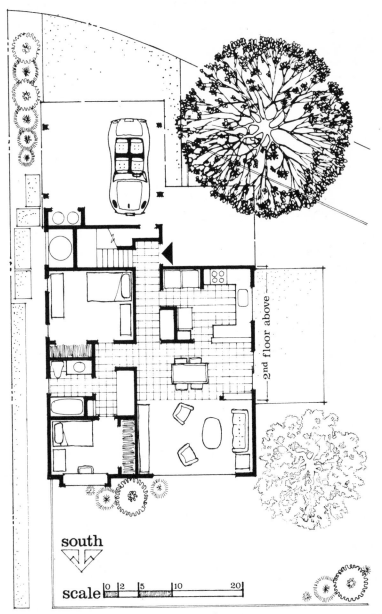

south

scale 0 2 5 10 20

The Watson Residence

Design: *John Hofacre*
Construction: *Mike Corbett*
Price Range: *(Adj. 1978)*
$64–68,000
Living Area: *1,608 sq. ft.*
Built: *1977*

Heat-circulating fireplace in living room provides backup heat.

A statement of John Watson's solar philosophy is made by this three-bedroom, two-bath home with separate den and sewing room. John helped in the construction of the beautiful all-oak kitchen cabinets. A dark brown tile floor used throughout adds richness. A cool north summer patio is surrounded on three sides by the U-shaped house; the open side enjoys a view over a Village vineyard. The garage, located on the north side, protects the south solar facade.

The System

The Watson house has 10 4' × 6' copper flat plate collectors in the 60° south roof. These supply both hot water and solar heating. The hot water is pumped from the collectors to a storage

Flat plate collectors lining the south roof soak up the sun.

tank. Another pump circulates the water through polybutylene plastic pipe spaced 9 in. apart in the concrete slab. The house also uses simple passive design with 175 sq. ft. of double-pane glass on the south and tile floors for storage. Cooling is by cross-ventilation with light exterior colors, overhangs, and eventual landscaping to help reduce cooling demand.

Performance

The house achieves around 70–80% solar heating and provides much of the hot water. The backup heating is provided by a 40-gallon gas water heater that heats the water used in the radiant slab, and a fireplace. This performance for cooling is similar to that for the simple passive house.

Improvements

The house would work better with insulated drapes and shutters on the windows. The plastic film used for inner glazing for the collectors should be removed or replaced with glass, to reduce glare and wrinkling.

The 60° panels do create glare for neighbors and should be redesigned at a different angle or less evident panel arrangement. Also, a vent at the top of the vaulted living room ceiling would allow the hot summer air to escape more easily.

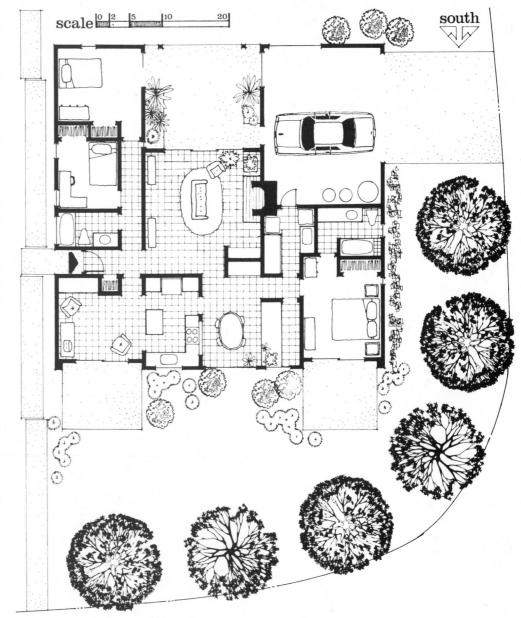

scale 0 2 5 10 20

south

The Solar Reflector Residence

Design: *John DeLapp, John Hofacre, Dave Springer*
Construction: *John DeLapp, Dave Springer*
Price Range: *(Adj. 1978) $66–70,000*
Living Area: *1,506 sq. ft.*
Built: *1978*

The Solar Reflector residence is one of the three buildings in Village Homes that has received federal grants for solar development. It has a very efficient floor plan with a master bedroom, study and bath upstairs, and two bedrooms and one bath downstairs. The clerestory windows with side panels provide abundant, soft, natural light to the entire second floor. An intimate deck is located off the master bedroom. Attractive canvas sunscreens have been added to the arbor for summertime shading of the tall south-facing windows.

The System

The Solar Reflector residence is one of the active houses that utilizes good passive design. A series of eight (24 sq. ft. copper with aluminum fin) collectors on the 60° second-story south wall receive direct sun and sun reflected from the reflective aluminum roof over the south part of the house. This configuration allows about 30% more light to fall on the collector without increasing the collector size, and thus the cost. Water is pumped from the collectors to a fiberglass storage tank. From there it is pumped to polybutylene plastic pipe embedded three in. deep in the concrete slab and in a two-in. lightweight slab on the second floor. The pumps are connected to a control system that operates them automatically. Much of the slab is

South windows provide passive heat collection.

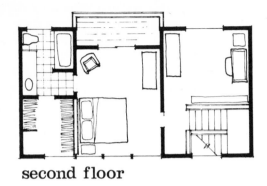

second floor

Clerestory lights up second floor.

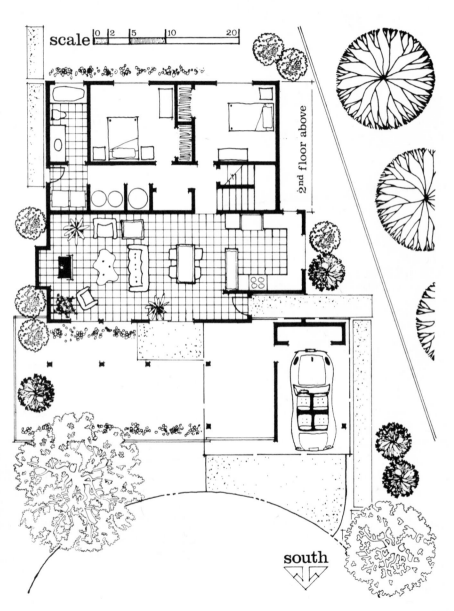

scale 0 2 5 10 20

2nd floor above

south

carpeted, but the south kitchen, dining, and living areas are tiled and will help passive heating and cooling. Cooling is by cross-ventilation aided by south overhangs, arbor, openable second-story windows, and minimal glass facing east and west.

Collector configuration helps hide the collectors from the street view, providing partial relief from any reflected glare.

Performance

The Solar Reflector house should meet about 75–85% of the heating demand for the house with the active and passive solar contributions. The system should also provide 60% of the hot wa-

ter. The house will be naturally cooled with a few warm evenings and afternoons.

Improvements

Insulated shutters and drapes are really the only additions that should be made. The tilted reflector might well be replaced with a flat roof. It would only reduce performance of the reflector a little and would be less expensive to build and less likely to leak.

The active system, where space

Canvas, which provides summer shade, can be removed in winter.

South roof reflects sun onto collectors.

Controls for solar system.

heating and water heating are combined, results in less hot water for domestic use in the winter. The temperature is reduced by space heating. The backup water heaters would best be located on the exterior since the air they draw for combustion increases infiltration.

Hard-to-reach clerestories open with a pole.

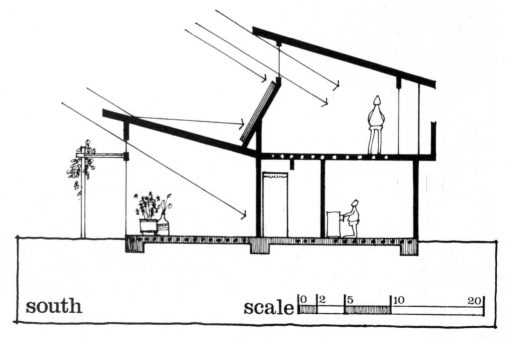

south   scale   0   2   5   10   20

Chapter Four

Special Features

Some of the houses in Village Homes have special features in addition to the basic design concepts and details and solar systems described in the previous sections. These include low energy and water use appliances, greywater drains, and solar clothes dryers. These are presented here because they are worth a closer look and deserve consideration in both new and existing houses.

Appliances

We have found through experience in Village Homes that energy bills for different families living in identically designed houses can vary enormously. Part of this is due to behavioral patterns, but a great deal is a result of the number or type of appliances used. The choice of appliances can be critical in keeping energy use down. The most critical choices come in selecting backup heating and cooling systems, hot water heating backup, refrigerator/freezer, clothes dryer, and stove.

Heating and Cooling

The most commonly used method of backup heating in Village Homes is the wood stove, either a Franklin or airtight stove. Both are more efficient than the more traditional fireplace. This is particularly true for an exterior fireplace because the R-value of an uninsulated brick fireplace is about the same as a window. In addition, the chimney acts as a heat stack that pumps heat out of the house both when it is burning (it may actually have more heat loss than heat gain) and when it is empty, unless the damper is very airtight. Use of a heatilator vent system and addition of doors can considerably improve the performance of a fireplace.

A Franklin stove that is well sealed can provide the aesthetics of open flames yet still be closed for reasonably efficient burning. It is a bit hard to generalize about efficiency of these wood stoves as it depends in part on manufacturer, installer, and use, but the rule of thumb is that a Franklin stove delivers 30% of the Btu content of the wood for use compared to about 10% for a good fireplace.

A much better choice, in fact the only real choice in areas colder than Davis, is the airtight stove in one of its many forms. Several types are in use in Village Homes including Morsø, Jøtul, Frontier, Riteway, etc., and the owners are justifiably proud of them. They offer much higher efficiency than a Franklin or a fireplace, up to 50%+, and therefore require less wood and burn longer with a given charge of wood.

The wood for these heaters has in large part been the scrap generated by the development. Even with fairly careful construction (not all the builders in Village Homes are very careful about waste), lots of tag ends and miscella-

Morsø.

Jøtul.

neous wood scraps are generated. The developer has established drop areas for this wood scrap in the development, and as a result the scrap is used rather than being buried in a sanitary (insanitary seems more appropriate) landfill. It has been interesting to watch these large piles of scrap disappear piece by piece into wheelbarrows, bags, and occasionally trucks.

This soft wood is satisfactory for use in Davis because not too many fires are needed in winter. In colder areas it would be useful as kindling for hardwoods, but the high tar content would make its use as the only fuel less satisfactory. In the future prunings from the orchards in and around the area will fill some of the demand.

Wood heating is not without drawbacks, however. On still, winter days with an inversion, the air quality deteriorates noticeably. More efficient stoves would reduce but not eliminate this.

Frontier.

Franklin.

The banks and city also required traditional backup heating systems, and the predominant choice for these has been gas. Because the houses are energy-efficient and have a high proportion of their heating demand met by solar energy, they can often use a very small wall heater. On most of the houses these are completely turned off all or almost all of the year. When they are never used, it is desirable to seal the vents. Setback thermostats can offer substantial savings with no change in comfort. They are used on a few of the houses and allow the furnace to keep the houses at 68°F during the day, yet only 62°F at night.

Several houses make more efficient use of their heating system, be it solar, wood, or gas, by including a small duct and fan to carry air from the ceiling back to the floor. These small fans, 1/10 - 1/100 hp, can keep cooler areas like an isolated north bedroom warm and help keep the loft or main room cooler.

The casablanca fans used in several houses perform a similar role in the win-

Jøtul.

Sierra.

ter. Summer, however, is when they are most useful, providing comfort by increasing air flow. The most commonly used fan has two speeds and uses only 40 watts on low speed.

A few houses in the development have air conditioners, but most rely only on natural cooling with the cool sea breeze. The people using air condition- ers have chosen the more efficient models. For good ventilation when the sea breeze is weak or for security ventilation without open windows and doors, exhaust fans are used with security vents. The fan blows warm air from the ceiling directly outside or through the attic and draws cool air through the vents and partially opened, secured windows.

Hot Water Heating

Some special features for backup water heating are also included in a few of the homes in Village Homes. Many of the houses have gas backup heaters. Others use either manual or automatic electric heaters. Added insulation on these backup heaters is often included.

Casablanca fan.

A more interesting backup heater that has just been installed relies on the time-tested use of a heat exchanger in the wood stove to heat the water.

Reducing the use of hot water also helps save energy, and the houses in Village Homes almost all have low water use appliances and water saver faucets and showers. In addition, a couple of houses use an instant electric hot water heater that heats water to 180°F in a well-insulated quart container right at the kitchen sink. These are controlled with a wall switch so they can be turned on only when needed for a cup of coffee, soup, or on a cold cloudy day, to heat up solar water enough for dishwashing. Use of these little heaters saves water because hot water doesn't have to be run until it reaches the sink. This saves many gallons of water over a month and considerable energy — energy that is used to heat water in the hot water heater to warm the pipe run and energy that would otherwise be used to heat water on the stove.

Refrigerators

The energy used in many of today's refrigerators is staggering. In a study in Davis conducted by Bill Kopper,[24] some homeowners were found to be paying more than $12 and using almost 500 kwh per month for refrigeration. An energy-efficient refrigerator without butter warmer, skin heater to prevent sweating, and automatic defroster can reduce energy use by as much as a factor of 10.

Even greater savings measures are possible and are being considered for future use at Village Homes. One of the easiest steps is relocating the heat exchangers to a cooler spot, either below the floor or outside. This allows the refrigerator to work with a much higher temperature gradient where it will be more efficient. Rather than trying to lose heat to air at 120°F it may be able to lose it to 60°F or less. Further savings could be realized by putting the heat exchanger in a "cool pool," an open pond of water fully shaded throughout the day.[25]

Ideally, all houses should have a built-in refrigerator with very good insulation (4-5 times present use) and a remote heat exchanger. With good seals and clever design, energy use could probably be dropped to 20 kwh per month. For bulk freezing of vegetables and fruits a community freezer makes much better sense than a series of inefficient home freezers. Localizing the freezing would provide a good local heat source for a vegetable dryer or assist hot water heating for a community laundry.

Cooking Stove

The use of energy for cooking is almost unnoticeable in a traditionally built house with gas heating and gas hot water, but in a solar house it may be one of the only uses of auxiliary energy. A pilotless stove can save 50% of the energy consumed by a gas stove with pilot. This is particularly desirable in the summer as it reduces the heat load on the building. A stove with a pilot can be almost as efficient and convenient if the pilot is turned off and a piezoelectric or glow starter is used to light the burners and oven.

Lighting

The use of energy for lighting is minor in most residences, but it is still worth reducing its use where possible. Quality of light is much more important than quantity, and clever use of task lights in work areas will provide safer and more energy-efficient lighting than bright area lights. The desirability of fluorescent lights is currently being questioned because of possible adverse health impacts, but they make good sense for areas where exposure will be limited.

Daylighting is best and can be maximized through careful placement of windows. Long thin windows near walls wash the walls with light and provide good diffuse lighting. Light-colored walls and ceilings greatly increase use of both natural and artificial light. Spaces such as storage cabinets, closets, and attics can often be naturally lighted with a small window, and this will make it less likely that lights will accidently be left on.

Outside lighting at night may be desirable, but is often overdone. The difference in visibility around the house with a 20-watt rather than a 100-watt outside light may be almost unnoticeable and yields a fivefold savings in energy. Experimentation on various houses has resulted in a reduction in energy use of 200–400% for exterior lights.

The street lighting in Village Homes is also energy-conserving. By careful political campaigning, the density of street lights was reduced to about one-half that of a standard development in Davis.

Greywater Drains

Provision has been made on a few of the houses for draining greywater from the shower or tub into the garden or onto shrubs. This was largely a spin-

Upstairs bathtubs are fitted with two drains: one to the garden, the other to the sewer.

An interior clothes drying rack.

off of the recent California drought when greywater saved countless plants and gardens from death. Despite the experience gained in the drought, the county health department has refused to allow the installation of these drains and has threatened prosecution unless the existing ones are permanently plugged. This battle is still in progress.

Solar Clothes Dryers

Many of the houses in Village Homes have solar clothes dryers — our

new name for the old-fashioned clothesline. Many houses in Village Homes also have indoor drying racks for use in the winter. The energy savings realized by not using the clothes dryer are quite substantial.

ANNUAL TOTAL NET ENERGY* COST: TWO LOADS PER WEEK[26]

Electric dryer	4.7 million Btu's
Gas dryer	2 million Btu's
Clothesline	62,000 Btu's

*including energy cost of manufacturing, etc.

Closing

Village Homes very clearly illustrates the substantial improvements that can be made in the patterns of development and building, working within the current institutional setting. Village Homes was constructed using conventional financing and building practices and conforming to current regulations, proving that it can be done.

The energy and dollar savings are, of course, important, but even more important is the increase in independence that a passive solar system and energy- and resource-efficient community can provide. A house that will work well even when the power is off can help provide peace of mind in an increasingly troubled world.

Village Homes is only a beginning. It represents one of the first steps toward a sustainable future. Just as houses within Village Homes have been improved as we have gained experience, we hope that future developments will be improved to become largely self-sufficient neighborhoods.

Most of the necessary techniques and equipment are now available to do this. The challenge is to combine these many simple, practical, and economical steps so they work together and will meet the often archaic and illogical development and building codes. This will be easiest to do in a new development, but retrofitting existing developments should also receive more extensive consideration along these lines. Our survival as an independent, free people and as a species may depend on it. Let us begin!

*A tree as great as a man's
embrace springs from
a small shoot;
A terrace nine stories high
begins with a pile
of earth;
A journey of a thousand miles
starts under one's
feet.*

Lao Tzu, sixth century B.C.
Tao Te-Ching (1972)
Vintage Press

References

1. Jan Hamrin (1978) *Two Energy Conserving Communities: Implications for Public Policy,* Doctoral Thesis, U.C. Davis.

2. See note 1 above.

3. Oscar Newman (1973) *Defensible Space,* Collier, New York, N.Y.

4. Jon Hammond et al. (1974) *A Strategy for Energy Conservation,* Living Systems, Winters, Calif.

5. David A. Bainbridge (1977) *Planning for Energy Conservation,* Living Systems, Winters, Calif.

6. Marshall Hunt and David A. Bainbridge (1978) "The Davis Experience," *Solar Age,* June.

7. James Ridgeway (1977) *The Davis Experiment,* The Elements, Washington, D.C.

8. Jon Hammond et al. (1977) *The Davis Energy Conservation Report,* Living Systems, Winters, Calif.

9. See note 8 above.

10. Marshall Hunt et al. (1975) *The Davis Energy Conservation Building Code,* Davis, Calif.

11. Len Myrup and Don Morgan (1972) *Numerical Model of the Urban Atmosphere: The City-Surface Interface,* U.C. Davis, Ag. Eng.

12. Carol and John Steinhart (1974) *Energy,* Duxbury Press, Scituate, Mass.

13. David A. Bainbridge and Marshall Hunt (1977) *The Effect of Roof Color and Material on Temperature,* Living Systems, Winters, Calif.

14. Bruce Anderson et al. (1977) *The Fuel Savers,* T.E.A., Harrisville, N.H.

15. Ken Butti and John Perlin (1977) "Solar Water Heaters in California, 1891-1930," *CoEvolution Quarterly,* Fall.

16. Murray Milne (1976) *Residential Water Conservation,* California Water Resources Center Report, #35.

17. F. A. Brooks (1936) *Solar Energy and Its Use for Heating Water in California,* U.C. Berkeley, Ag. Exp. Station Bulletin #602.

18. David A. Bainbridge and Jeff Reiss (1978) "Breadbox Designs," *Alternative Sources of Energy,* #34.

19. Bruce M. Anderson with Michael Riordan (1976) *The Solar Home Book,* Cheshire Books, Harrisville, N.H.

20. Bruce Maeda et al. (1978) *A Simple Solar Home for California,* Solar Office, California Energy Commission, Sacramento, Calif.

21. Doug Balcomb, J. C. Hedstrom, and Robert McFarland (1977) "Thermal Storage Walls for Passive Solar Heating Evaluated," *Solar Age,* August.

22. Gary Starr et al. (1977) *Resource Conservation for Residential Buildings,* Final Report N.S.F. Grant #77-05558, Davis, Calif.

23. Bill and Susan Yanda (1976) *An Attached Solar Greenhouse,* The Lightning Tree, Santa Fe, N.Mex.

24. Bill Kopper (1976) *Energy Conservation Advisory Program,* Living Systems, Winters, Calif.

25. David A. Bainbridge (1978) "The Indio Cool Pool Experiment," *Alternative Sources of Energy,* #32.

26. See note 5 above.

Glossary

active system: a solar system which relies on external power to work, typically for pumps, fans, and controls.

airtight stove: a stove with very well sealed joints, allowing very precise control of the draft and high efficiency.

altitude: the elevation of the sun above the horizon.

azimuth: the angle of the sun from south.

backup system: a system used to augment a solar system.

breadbox: a solar water heater where collection and storage are combined.

Btu: British thermal unit, the amount of heat needed to raise the temperature of a pound of water 1°F.

calorie: the amount of heat needed to raise the temperature of a gram of water 1°C.

clerestory: a window inserted between a lower roof and a higher roof.

collector: a device for gathering the sun's energy — as simple as a south-facing window or as complicated as a pumped concentrating collector sitting on the roof.

collector efficiency: a description of the amount of energy gathered by a collector compared to the amount striking it.

comfort range: the range of climatic conditions within which people feel comfortable; this depends on the clothing, age, health, activity, and attitude of the people involved.

conduction: the transfer of energy from atom to atom or molecule to molecule.

coolth: energy associated with motion of molecules below the energy level of surrounding areas.

degree-day: a deviation of 1°F from a reference temperature, usually 65°F, for a day. Both heating degree-day and cooling degree-day are used.

diffuse radiation: solar radiation scattered by dust, water vapor, or gasses in the atmosphere.

direct radiation: solar radiation received directly from the sun.

efficiency: a description of the ratio of energy output for a given input, used for collector, system, stove, and house descriptions.

flat plate: a collector where solar energy is collected on a flat surface, without reflectors or focusing lenses.

forced ventilation: mechanically aided ventilation using some sort of fan or blower.

heat: energy associated with the motion of atoms or molecules at or above the energy level of surrounding areas.

heat capacity: the description of a material's ability to store heat; the amount of energy needed to raise the temperature of a cubic foot of the material 1°F.

heat exchanger: a device used to transfer heat from one fluid to another, air-liquid, liquid-liquid, or air-air. These must be double-walled when antifreeze is in one system and potable water in the other.

hybrid system: a system that combines both active and passive features.

induced ventilation: ventilation using the difference in density or air at different temperatures to cause the air flow.

infiltration: the movement of air into a house through cracks around windows, doors, seams, utilities, and from the opening of doors and windows. It can account for as much as one-half of the heat loss or gain in a well-insulated house.

insulation: materials used to reduce heat flow, typically fiberglass or foam.

internal heat gain: heat generated in the house from appliances, lights, people, and pets.

longwave radiation: longwave or thermal radiation is associated with low temperature, radiant transfer.

mass: see thermal mass.

natural ventilation: ventilation using natural breezes.

passive system: a solar system that will work even if the power goes off. It will typically use elements of the building to collect, store, and circulate energy. Collection and storage are commonly combined rather than separate so no pumps, fans, or controls are needed to transfer energy from point of collection to point of use.

percent solar: a crude measure of the amount of heating or cooling done by a solar system compared to the total demand. As used here, it includes internal heat gain as part of the passive system. The percent will vary considerably with different families and the temperature variation they allow. A house that gets 90% heating with a range from 60°-80°F may get only 60% between 65°-75°F.

price range: price range was used in this book rather than actual cost because it is difficult to figure exact costs and have them comparable with different contracts and builders. Price range is a little better. Cost increases have been remarkable in the last three years. Current rate of price increase in construction materials is 24% a year. Lots in Village Homes have doubled in price in three years due to inflation and increased city fees.

radiant temperature: the surface temperature of walls, ceiling, floor, or thermal mass.

radiation: flow of energy by electromagnetic waves (see longwave radiation, solar radiation).

reflected radiation: solar radiation reflected by light-colored or polished surfaces — either natural surface or building components. Can be used to double solar gain.

resistance or R-value: a measure of a material's ability to resist or retard heat transfer.

retrofit: to add a solar system or component to an existing structure.

security ventilation: a system of natural, induced, and forced ventilation that can be used without open windows or doors.

shading: the obstruction of the sun by an overhang, arbor, wall or landscaping. The choice of trees for shading a south wall is critical because a thick tree may block one-half of the sun even with its leaves gone.

shutters: rigid, insulated panels used to reduce heat loss or gain through a window or skylight. They must seal very tightly around the edges.

skylight: a window set in the roof.

solar access: portion of the sky needed to ensure enough sunlight is received to make a solar system work.

solar control: use of shading, colors, and materials to reduce unwanted heat gain.

solar radiation: electromagnetic radiation emitted by the sun, with most of the energy concentrated in the shortwave portion of the spectrum. The peak is near 0.5 microns.

solar rights: legal protection of solar access.

south: true south rather than magnetic south.

specific heat: the amount of heat needed to raise the temperature of a pound of material 1°F.

sun tempered: a standard dwelling that uses good solar orientation to meet much of its heating demand.

thermal mass: the use of materials with high heat capacity to store energy in a structure. This mass acts as a battery to solar heat or natural coolth.

thermal storage: the amount of thermal mass in a building determines its thermal storage.

thermosiphon: a solar system that places the collector lower than the storage so that the less dense hot water will rise naturally to storage. In a thermosiphon system no pump is necessary to move the water from collector to storage and around again. This arrangement can also be used for freeze protection. Reverse thermosiphon will occur when water nears freezing and becomes a little less dense.

tilt angle: the angle of a collector relative to the ground. The best angle for winter heating is the latitude $+10°$, while the best for year-round heating is the latitude $-10°$. However, even varying the collector tilt considerably more will not affect performance very much.

water-filled culvert: a culvert or large pipe, sealed at the bottom and filled with water, that is used for thermal mass.

Appendix A

SOLAR RIGHTS
AMENDMENT NO. 3 TO DECLARATION OF
COVENANTS, CONDITIONS AND RESTRICTIONS

Whereas, that certain Declaration of Covenants, Conditions and Restrictions was recorded on October 31, 1975, as instrument No. 15574, in Book 1166, Page 385 of official records,

Now, therefore, the lot owners, and the declarant, Village Homes, do hereby modify said Declaration of Covenants, Conditions and Restrictions as follows:

SOLAR RIGHTS

All south-facing glass and solar space heating collectors in each house shall remain unshaded from December 21 to February 21 between the hours of 10 A.M. and 2 P.M. (solar time), except as provided herein.

All roof-top solar hot water collectors on each house shall remain unshaded each day of the year between the hours of 10 A.M. and 2 P.M. (solar time), except as provided herein.

Shading caused by the branches of deciduous trees shall be exempt from this restriction.

Shading caused by original house construction, or fences built within six (6) months of occupancy shall be exempted from this restriction only upon special approval of the Village Homes Design Review Board.

Homeowners may encroach upon their own solar rights.

The Board of Directors of the Village Homeowners' Association shall have the authority to enforce this restriction.

Appendix B

SOLAR WATER HEATING EXPERIENCE
IN VILLAGE HOMES; AND THREE SETS OF
GUIDELINES FOR WATER HEATING INSTALLATIONS

The rebirth of the solar industry in California coincided with the development of Village Homes. Initially there was little or no information available on the design of solar water heaters. One of the best references on the subject was a technical paper written in the 1930s by F. A. Brooks of the University of California. Consequently, many of the solar water heaters in Village Homes were experimental prototypes. The advantages and disadvantages of the different designs quickly became apparent. This section describes some of the lessons that were learned.

Several thermosiphon water heaters were installed by Natural Heating Systems before the inception of Village Homes, and experience with these was good. Therefore, the use of these was carried on in Village Homes. Initially the only tanks available were constructed of galvanized steel. As soon as they became available, glass-lined steel tanks were used. Although early solar water heaters in California used galvanized tanks successfully, the efficiency of collectors at that time was not as high as it is now with better insulating materials. The water probably gets much hotter with modern collectors, and at the time this was written, three galvanized steel tanks had failed, one of them in Village Homes. It is apparent now that galvanized steel tanks are to be avoided at all cost. The damage a corroded tank can cause is substantial. Glass-lined steel tanks should be equipped with anode rods to further inhibit corrosion. For thermosiphon systems the freeze protection using the reverse thermosiphon seems to work very well in this climate (30°F design temperature). Temperatures as low as 28°F have been experienced with no difficulties. However, we recommend that the collectors be drained if the temperature is expected to drop below 28°F.

A few of the houses in Village Homes could not accommodate an attic storage tank, and therefore a thermosiphon system could not be employed. In order to conserve materials' costs, single-tank pumped-flow solar water heaters were initially used. These circulated potable water directly through the collectors and had an electric heating element in the storage tank for backup heating. After experiencing several high utility bills, it became apparent that the pump was causing the stratified electrically heated water at the top of the tanks to mix with the cooler water below. Thus the effectiveness of the solar collectors for heating the water was drastically reduced. These systems were converted to two-tank systems using gas or electric backup water heaters. A few single-tank thermosiphon systems were tried and experiences were similar. If a baffle or diffuser were installed in the tank to minimize mixing, then single-tank "open" systems might be practical. Single-tank closed-loop systems are immune to this problem.

All of the water heating systems in Village Homes are equipped with valves that enable the backup water heater to be bypassed and shut off to minimize heat loss. The amount of gas that is used by a water heater in simply keeping the water hot is sufficient to heat 45 gallons of water from 60°-140°F per day. Thus, the energy saved by turning off backup water heaters during the summer can be substantial. Most of the Village Home residents rely totally on solar heating for hot water between May and September. This alone results in a savings of over 6,700 gallons of hot water per household per year, not counting what is produced by the solar water heater.

Ideally, the backup water heater on a solar heating system should have a very small storage capacity, a high recovery rate, and a high efficiency. Instantaneous water heaters that meet these criteria are on the market, but they are not designed to operate with solar preheated water. Rather than being actuated by temperature, they are actuated by flow. We hope the manufacturing industry will respond to the needs of the solar industry by providing the materials that will help make solar water heating even more successful.

Guidelines for Installing Breadbox Hot Water Heaters
by Dave Bainbridge

1. *Obtain needed building permits.*
2. *Face collector due south (insulate box well, six–eight in. of insulation) in an area that can support the weight of the*

*tanks. Total system weight may be 1,200 lbs., so extra
bracing may be required.*

3. *Paint surfaces black (reflectors on sides and back where
   bounce back is minimal will help).*
4. *Mount tanks vertically, not horizontally.*
5. *Incline collector surface and tanks at latitude angle (+ 10° if
   used in winter).*
6. *Plumb tanks in series, with last tank being the feed tank
   (cold water in bottom, hot drawn off top).*
7. *Use long, thin tanks if available (mobile home tanks are
   good).*
8. *Size tanks to provide about 20–30 gallons per day per
   person (typically three 30-gallon tanks are used, which will
   provide for three to four people).*
9. *Size collector to provide at least 1 sq. ft. of glazing per 2.5
   gallons.*
10. *Provide insulated shutters or drapes and reflectors to
    improve performance.*
11. *Carefully insulate exposed pipes to prevent freezing (tanks
    are safe except in fairly cold areas where they should be
    drained in winter).*

Guidelines for Installing Thermosiphon Water Heating Systems by Dave Springer

1. *Obtain needed building permits and approvals.*
2. *The collectors should be mounted facing due south (± 15°)
   and tilted at an angle of from 15°–45° (an angle equal to the
   latitude is optimum but lower angles will still work well).*
3. *The collectors must be unshaded between the hours of 9 A.M.
   and 4 P.M. at any time of the year.*
4. *The collectors and storage tank should be placed as close to
   the conventional water heater as possible and connecting
   plumbing should be well insulated.*
5. *The vertical distance from the top manifolds of the collectors
   to the bottom of the storage tank must be 1 ft. ± 3 in. for
   optimum freeze protection and system performance.*
6. *Piping between the collectors and storage tank must be*

sloped to allow any air in the system to bubble into the storage tank without being trapped in the plumbing. This is critical to ensuring that the water will circulate. The piping should be well insulated.

7. *If the storage tank is to be installed in the attic space, the roof pitch must be a minimum of 4 in 12 and the horizontal distance from the south wall of the house to the peak of the roof must be at least 15 ft. If the roof pitch is at least 6 in 12, the storage tank may be mounted vertically within the attic. Otherwise, a special enclosure must be constructed for the tank.*

8. *The collectors should be equipped with the appropriate plumbing to permit draining during extremely cold (28°F) weather.*

9. *The storage tank should be connected to the conventional water heater so that the water heater can be bypassed during summer months to avoid loss of heat (about 33,000 Btu's per day for gas water heaters).*

10. *It has been our experience that users may not always drain their collectors during a freeze. We therefore recommend the addition of a Freezeton Valve (Sarco Industries, 1457 Rollins Road, Burlingame, CA 94010). This valve automatically trickles water from the collector during freezing conditions. In climates where the temperature drops below 32°F more than 10 times per year, a simple thermosiphon system should not be used.*

Guidelines for Installing Pumped Flow Water Heaters by Dave Springer

1. *Same as thermosiphon.*
2. *Same as thermosiphon.*
3. *Same as thermosiphon.*
4. *Same as thermosiphon.*
5. *Systems circulating antifreeze in a closed-loop circuit with a heat exchanger for heating the potable water are less efficient and more expensive, but are more dependable if used in areas where freezing is frequent.*

6. *Systems circulating potable water through the collectors are the systems of choice in regions with winter design temperatures warmer than 30°F. However, careful attention to freeze protection must be given. Control systems that drain the collector are safer to use than systems relying on recirculation for freeze protection since the latter type does not provide protection during power failures unless special provisions are made.*
7. *Solar water heaters circulating potable water should use a separate water heater for backup heating (two-tank system), whereas the closed-loop, antifreeze system may function well as a one-tank system, which employs a heating element near the top of the solar storage tank for backup heating.*

Appendix C

ACCESS TO MATERIALS AND EQUIPMENT

Casablanca Fans:

W. W. Grainger, Inc.
5959 West Howard Street
Chicago, IL 60648
Also see advertisements in *Sunset* magazine, and mail order catalogs from Montgomery Wards and Sears & Roebuck.

Small Fans and Blowers:

W. W. Grainger, Montgomery Wards, and Sears & Roebuck

Security Ventilation — sliding glass doors with lockable vent position:

Blomberg Window Systems
1453 Blair Avenue
Sacramento, CA 95822

Shutters:

ThermaRoll Corporation
512 Orchard Street
Golden, CO 80401

Shutters, Inc.
115 West 23rd Street
Hastings, MN 55033

Roto International
P.O. Box 73
Essex, CT 06426

Amrol Exterior Rolling Shutters
Pease Company
New Castle, IN 47362

Drapes:

Appropriate Technology Corporation
P.O. Box 975
Brattleboro, VT 05301

IS Company
17 Water Street
Guilford, CT 06437

Thermal Technology Corporation
Box 130
Snowmass, CO 81654

Drape Material:

Foylon — reflective material
Duracote
350 North Diamond
Ravenna, OH 44266

Thinsulate — insulating mesh
2223-6SW
3M Center
Saint Paul, MN 55101

Dacron Holofil II — insulating mesh
E.I. DuPont DeNemours and Co.
Wilmington, DE 19898

Metal Shadescreen:

Koolshade Corporation
722 Genevieve Street
Solana Beach, CA 92075

Kaiser Shadescreen
ASC Distributing Inc.
1645 West Buckeye
Phoenix, AZ 85007

Fiberglass Shadescreen:

Chicopee Mfg. Co.
Cornelia, GA 30531

Phifer Wire Products, Inc.
Tuscaloosa, AL 35401

Phifer Shadescreen
14408 East Nelson
City of Industry, CA 91744

J. P. Stevens & Co. Inc.
Walterboro, SC 29488
and
1185 Avenue of Americas
New York, NY 10036

Owens-Corning Fiberglas
Granville, OH 43023

PPG Industries, Inc.
Fiberglass Division
Pittsburgh,PA l5222

Vinco
P.O. Box 212
Laurel, VA 23060

Storage Tanks and Containers:

Kalwall Corporation
Solar Components
P.O. Box 237
Manchester, NH 03103

One Design, Inc.
Mountain Falls Route
Winchester, VA 22601

Tabline Company
P.O. Box 1135
Berkeley, CA 94701

ACT Polydrum (5 gal.)
Advanced Chemical Technology
City of Industry, CA 91744

ARMCO Steel Helicor Culverts
ARMCO Steel
Road 32
Davis, CA 95616

Montgomery Wards
Fuel Oil Storage Tank
(Part #81 C49988F in Fall and
Winter 1978 Catalog)

(Army Surplus Cordite Cans are
also excellent.)

Water Heater for Wood Stoves:

Blazing Showers
P.O. Box 377
Point Arena, CA 95468

Catalogs and Accessories to Tools:

Soft Tech (1978) $5
Jay Baldwin and Stewart Brand
CoEvolution Quarterly
P.O. Box 428
Sausalito, CA 94965

Solar Age Catalog (1977) $8.50
Bruce Anderson and Sandra Oddo
SolarVision
Harrisville, NH 03450

The Passive Solar Catalog
(1978) $5 + $1 handling
David A. Bainbridge
The Passive Solar Institute
P.O. Box 722
Davis, CA 95616

Appendix D

RESOURCES

Quite often the resource lists with this type of book are so long and intimidating that they effectively stop the reader from pursuing any subject further. This list is highly selective and includes only those sources we feel are excellent.

Agriculture

J. Jeavons (1974) *How to Grow More Vegetables,* Ecology Action, Palo Alto, Calif.

Richard Merrill, ed. (1976) *Radical Agriculture,* Harper & Row, New York, N.Y.

J. I. Rodale (1978) *The Encyclopedia of Organic Gardening,* revised edition, Rodale Press, Emmaus, Pa.

Appropriate Development

R. J. Congdon, ed. (1977) *Introduction to Appropriate Technology,* Rodale Press, Emmaus, Pa.

Amory Lovins (1977) *Soft Energy Paths,* Ballinger, Cambridge, Mass.

Michael North, ed. (1977) *Time Running Out?,* Universe Books, New York, N.Y.

E. F. Schumacher (1975) *Small Is Beautiful,* Harper & Row, New York, N.Y.

Climate and Microclimate

Rudolf Geiger (1965) *Climate Near the Ground,* 4th ed., Harvard University Press, Cambridge, Mass.

Baruch Givoni (1969) *Man, Climate, and Architecture,* Elsevier, New York, N.Y.

Victor V. Olgyay (1963) *Design with Climate,* Princeton University Press, Princeton, N.J.

Comfort

David Egan (1972) *Concepts of Thermal Comfort,* Tulane University of Architecture, Tulane, La.

P. O. Fanger (1973) *Thermal Comfort,* McGraw-Hill, New York, N.Y.

Baruch Givoni (1969) *Man, Climate, and Architecture,* Elsevier, New York, N.Y.

Community

Wendell Berry (1976) *The Unsettling of America,* Sierra Club Books, San Francisco, Calif.

Jane Jacobs (1961) *The Death and Life of the Great American Cities,* Vintage Books, New York, N.Y.

Oscar Newman (1973) *Defensible Space,* Collier, New York, N.Y.

Moshe Safdie (1974) *For Everyone a Garden,* MIT Press, Cambridge, Mass.

Energy Conservation

B.H.K.R. Associates (1977) *Minimum Energy Use Dwelling Handbook,* Department of Energy, National Technical Information Service, Springfield, Va.

Stewart Byrne (1975) *The Arkansas Story,* Owens-Corning, Toledo, Ohio.

Jon Hammond et al. (1975) *The Davis Energy Conservation Code,* Davis, Calif.

Marshall Hunt et al. (1976) *The Davis Energy Conservation Building Code Workbook,* Davis, Calif.

Steven Robinson and Fred S. Dubin (1978) *The Energy Efficient Home,* New American Library, New York, N.Y.

House Building and Design

L. O. Anderson (1973) *How to Build a Wood-Frame House,* Dover, New York, N.Y.

Etienne Grandjean (1974) *Ergonomics of the Home,* Halsted Press, New York, N.Y.

Barbara Kern and Ken Kern (1977) *The Owner Built Homestead,* Charles Scribner's Sons, New York, N.Y.

Ken Kern (1975) *The Owner-Built Home,* Charles Scribner's Sons, New York, N.Y.

Landscape

Gary O. Robinette (1972) *Plants, People and Environmental Quality,* U.S. Government Printing Office, Washington, DC 20402, stock no. 2405-0479.

Gary O. Robinette (1977) *Landscape Planning for Energy Conservation,* Environmental Design Press, Reston, Va.

R. F. White (1945) *Effects of Landscape Development on Natural Ventilation,* R.R. #45, Texas Engineering Experiment Station, Austin, Tex.

Planning

David A. Bainbridge (1978) *Solar Access: A Local Responsibility,* Solar Office, California Energy Commission, Sacramento, Calif.

Clare C. Cooper (1975) *Easter Hill Village,* Free Press, New York, N.Y.

G. C. S. Curtis (1975) *A Design Guide for Residential Areas,* Essex County Council, Tiptree, England.

Robert Sommer (1972) *Design Awareness,* Holt, Rinehart & Winston, New York, N.Y.

Undercurrents editors (1976) *Radical Technology,* Pantheon, New York, N.Y.

Peter van Dresser (1976) *A Landscape for Humans,* The Lightning Tree, Santa Fe, N.Mex.

References

American Society of Heating, Refrigerating and Airconditioning Engineers (1977) *Handbook of Fundamentals,* ASHRAE, New York, N.Y.

Environmental Science Services Administration (1968) *Climatic Atlas of the United States,* U.S. Department of Commerce, Washington, D.C.

Libby-Owens-Ford (1977) *Sun Angle Calculator,* Toledo, Ohio.

Charles G. Ramsey and Harold R. Sleeper (1972) *Architectural Graphic Standards,* John Wiley & Sons, New York, N.Y.

Clifford Strock and Richard L. Koral, eds. (1965) *Handbook of Air Conditioning, Heating, and Ventilating,* 2d ed., Industrial Press, New York, N.Y.

Services

Jim Leckie et al., eds. (1975) *Other Homes and Garbage,* Sierra Club Books, San Francisco, Calif.

Witold Rybczynski (1974) *The Ecol Operation,* McGill University, Montreal, Canada.

Carol H. Stoner, ed. (1977) *Goodbye to the Flush Toilet,* Rodale Press, Emmaus, Pa.

Brenda Vale and Robert Vale (1976) *The Autonomous House,* Universe Books, New York, N.Y.

Solar Home Design

Bruce M. Anderson with Michael Riordan (1976) *The Solar Home Book,* Cheshire Books, Harrisville, N.H.

David A. Bainbridge (1978) *The Passive Solar Catalog,* The Passive Solar Institute, Davis, Calif.

Charles Barnaby et al. (1978) *Solar for Your Present Home,* California Energy Commission, Sacramento, Calif.

James C. McCullagh, ed. (1978) *The Solar Greenhouse Book,* Rodale Press, Emmaus, Pa.

Solar Age editors (1977) *Solar Age Catalog,* SolarVision, Church Hill, Harrisville, N.H.

Sunset editors (1978) *Solar Heating Your Home,* Lane Publishing Co., Menlo Park, Calif.

Peter van Dresser (1977) *Homegrown Sundwellings,* The Lightning Tree, Santa Fe, N.Mex.

Solar Water Heating

Bruce M. Anderson with Michael Riordan (1976) *The Solar Home Book,* Cheshire Books, Harrisville, N.H.

F. A. Brooks (1936) *Solar Energy and Its Use for Heating Water in California,* UC Berkeley, Agriculture Experiment Station, Berkeley, Calif., from O.A.T., 1400 Tenth, Sacramento, Calif.

Transportation

Ivan Illich (1974) *Energy and Equity,* Harper & Row, New York, N.Y.

Robert Kirby et al. (1974) *Paratransit,* The Urban Institute, Washington, D.C.

Public Works Department Staff (1975) "Development of Bicycle and Street Facilities in Davis," Davis, Calif.